Great Minds

Also by the Authors

IH, *Buried Glory: Portraits of Soviet Scientists* (Oxford University Press, 2013)

IH, *Drive and Curiosity: What Fuels the Passion for Science* (Prometheus, 2011)

IH, *Judging Edward Teller: A Closer Look at One of the Most Influential Scientists of the Twentieth Century* (Prometheus, 2010)

MH and IH, *Symmetry through the Eyes of a Chemist*, 3rd edition (Springer, 2009; 2010)

MH and IH, *Visual Symmetry* (World Scientific, 2009)

IH, *The DNA Doctor: Candid Conversations with James D. Watson* (World Scientific, 2007)

IH, *The Martians of Science: Five Physicists Who Changed the Twentieth Century* (Oxford University Press, 2006; 2008)

IH, *Our Lives: Encounters of a Scientist* (Akadémiai Kiadó, 2004)

IH, *The Road to Stockholm: Nobel Prizes, Science, and Scientists* (Oxford University Press, 2002; 2003)

IH, MH, and BH, *Candid Science I–VI: Conversations with Famous Scientists* (Imperial College Press, 2000–2006)

IH and MH, *In Our Own Image: Personal Symmetry in Discovery* (Plenum/Kluwer, 2000; Springer, 2012)

IH and MH, *Symmetry: A Unifying Concept* (Shelter Publications, 1994)

R. J. Gillespie and IH, *The VSEPR Model of Molecular Geometry* (Allyn & Bacon, 1991; Dover Publications, 2012)

Great Minds

Reflections of 111 Top Scientists

Balazs Hargittai
Magdolna Hargittai
Istvan Hargittai

OXFORD
UNIVERSITY PRESS

Oxford University Press is a department of the University of
Oxford. It furthers the University's objective of excellence in research,
scholarship, and education by publishing worldwide.

Oxford   New York
Auckland   Cape Town   Dar es Salaam   Hong Kong   Karachi
Kuala Lumpur   Madrid   Melbourne   Mexico City   Nairobi
New Delhi   Shanghai   Taipei   Toronto

With offices in
Argentina   Austria   Brazil   Chile   Czech Republic   France   Greece
Guatemala   Hungary   Italy   Japan   Poland   Portugal   Singapore
South Korea   Switzerland   Thailand   Turkey   Ukraine   Vietnam

Oxford is a registered trademark of Oxford University Press
in the UK and certain other countries.

Published in the United States of America by
Oxford University Press
198 Madison Avenue, New York, NY 10016

© Oxford University Press 2014

All rights reserved. No part of this publication may be reproduced, stored in
a retrieval system, or transmitted, in any form or by any means, without the prior
permission in writing of Oxford University Press, or as expressly permitted by law,
by license, or under terms agreed with the appropriate reproduction rights organization.
Inquiries concerning reproduction outside the scope of the above should be sent to the
Rights Department, Oxford University Press, at the address above.

You must not circulate this work in any other form
and you must impose this same condition on any acquirer.

Library of Congress Cataloging-in-Publication Data
Hargittai, Balazs.
Great minds : reflections of 111 top scientists / Balazs Hargittai,
Magdolna Hargittai, and Istvan Hargittai.
pages cm
Includes bibliographical references and index.
ISBN 978-0-19-933617-3 (alk. paper)
1. Scientists—Interviews.   2. Scientists—Psychology.
I. Hargittai, Magdolna.   II. Hargittai, Istvan.   III. Title.
Q141.H2627 2014
509.2'2—dc23
2013040837

1 3 5 7 9 8 6 4 2
Printed in the United States of America
on acid-free paper

For Eszter, Michele, Matthew, and Stephanie

CONTENTS

Preface xi

Section One: Physicists
Zhores I. ALFEROV 3
Philip W. ANDERSON 6
Jocelyn BELL BURNELL 9
Catherine BRÉCHIGNAC 13
John H. CONWAY 16
Mildred S. DRESSELHAUS 19
Freeman J. DYSON 23
Jerome I. FRIEDMAN 26
Richard L. GARWIN 30
Vitaly L. GINZBURG 34
Donald A. GLASER 38
Maurice GOLDHABER 41
David GROSS 45
Antony HEWISH 48
Gerardus 't HOOFT 50
Wolfgang KETTERLE 53
Nicholas KURTI 56
Benoit B. MANDELBROT 59
Rudolf MÖSSBAUER 62
Yuval NE'EMAN 65
Mark OLIPHANT 69
Wolfgang K. H. PANOFSKY 72
Roger PENROSE 75
Arno A. PENZIAS 78
John POLKINGHORNE 81
David E. PRITCHARD 84
Norman F. RAMSEY 87
Vera C. RUBIN 90
Dan SHECHTMAN 93
Valentine L. TELEGDI 96
Edward TELLER 99
Charles H. TOWNES 102

Martinus J. G. VELTMAN 106
Steven L. WEINBERG 109
John A. WHEELER 112
Frank WILCZEK 116
Kenneth G. WILSON 119

Section Two: Chemists
Herbert C. BROWN 125
Erwin CHARGAFF 129
Mildred COHN 132
John W. CORNFORTH 135
Donald J. CRAM 138
Paul J. CRUTZEN 141
Johann DEISENHOFER 143
Carl DJERASSI 146
Gertrude B. ELION 149
Albert ESCHENMOSER 152
Kenichi FUKUI 155
Elena GALPERN 158
Darleane C. HOFFMAN 161
Roald HOFFMANN 165
Isabella L. KARLE 168
Jerome KARLE 171
Reiko KURODA 175
Yuan Tseh LEE 178
Jean-Marie LEHN 181
William N. LIPSCOMB 184
Stephen MASON 187
Bruce MERRIFIELD 190
George A. OLAH 193
Linus PAULING 196
John C. POLANYI 199
John A. POPLE 202
George PORTER 205
Vladimir PRELOG 207
F. Sherwood ROWLAND 210
Frederick SANGER 213
Glenn T. SEABORG 217
Nikolai N. SEMENOV 221
Frank H. WESTHEIMER 224
Ada YONATH 227
Richard N. ZARE 230
Ahmed H. ZEWAIL 234

Section Three: Biomedical Scientists

Werner ARBER 239
David BALTIMORE 242
Seymour BENZER 245
Paul BERG 247
Baruch S. BLUMBERG 250
Sydney BRENNER 253
Arvid CARLSSON 256
Aaron CIECHANOVER 260
Francis H. C. CRICK 263
D. Carleton GAJDUSEK 266
Walter GILBERT 270
Avram HERSHKO 272
Oleh HORNYKIEWICZ 275
François JACOB 279
Aaron KLUG 282
Arthur KORNBERG 286
Paul C. LAUTERBUR 289
Joshua LEDERBERG 292
Rita LEVI-MONTALCINI 295
Edward B. LEWIS 299
Peter MANSFIELD 302
Maclyn McCARTY 305
Matthew MESELSON 307
César MILSTEIN 310
Salvador MONCADA 313
Benno MÜLLER-HILL 316
Paul NURSE 319
Christiane NÜSSLEIN-VOLHARD 322
Max F. PERUTZ 325
Frederick C. ROBBINS 328
Jens Chr. SKOU 331
Gunther S. STENT 334
John E. SULSTON 337
Harold E. VARMUS 340
Alex VARSHAVSKY 343
James D. WATSON 346
Charles WEISSMANN 350
Rosalyn S. YALOW 353

List of Names 357
Index 369

PREFACE

Over the years, we have recorded and published hundreds of conversations with leading scientists, among them many Nobel laureates in physics, chemistry, materials science, and the biomedical sciences. For this book, we present excerpts from 111 of these conversations for a general readership. Our goal is to provide a glimpse into the thoughts of the leading scientists of the past half century and, in doing so, to humanize them.

The complete conversations appeared in a variety of magazines and journals, and most were collected in the six-volume, nearly 4,000-page *Candid Science* book series.[1] Those in-depth interviews provided detailed information about the lives and oeuvres, as well as the views and aspirations, of the participating scientists. But the level of scientific detail somewhat limited their readership to professionals in the various fields. We regretted this, as most of the conversations covered wide-ranging topics of universal interest as well. This led to the idea of compiling excerpts from a selection of these interviews in a single volume.

Science has never been more important to the health of society than it is today. On the one hand, more and more hard decisions have to rely on advances in modern science. On the other hand, it is becoming increasingly difficult for nonspecialists to grasp those advances. Even the best scientists are nonspecialists outside of their own area of inquiry. In a democratic society, political decision-making implies broad public participation through representation by elected officials. It is therefore desirable that the general public gain some knowledge about the progress being made in science and about the men and women who are responsible for this progress. Our hope is that our book will facilitate the acquisition of this knowledge in a pleasant, even entertaining way.

We are often asked how we were able to conduct hundreds of interviews with some of the most significant scientists of our time. Some have supposed that we received a huge grant; others, that there was a powerful institution behind our project. It was nothing like that. We used our own modest earnings as scientists to cover our expenses; and we seized every opportunity to meet famous colleagues that came our way. These opportunities came about when we taught and did research in various countries, were invited to international meetings and to departmental seminars, when distinguished scientists visited our hometown of Budapest, and even when we were on family vacations. Our own curiosity drove us to conduct these interviews and pursue answers to our questions.

---

[1] *Candid Science*, Vols. I–VI (London: Imperial College Press, 2000–2006).

The return was tremendous. We were learning from the best minds, the greatest experts, the discoverers themselves. Perhaps our interviewees sensed that their fellow-scientist interviewers were genuinely interested in learning from them, rather than merely carrying out an editor's assignment or meeting a deadline. Most of them opened up to us; and most left in the finalized text even some things they originally may have not meant to share with the public.

Our approach to these interviews was as follows: we recorded the conversations, transcribed them ourselves, and then returned the slightly edited transcripts to the interviewees for review. In contrast with journalistic practices, we did not ask our interviewees to limit their corrections to errors and obvious misunderstandings. Our interviewees could delete, rewrite, and add any amount of text without restriction. Some did, but most introduced only a small number of changes.

Some of our interviewees whom we had only known from the literature became our friends, and we have kept in touch with them. This was an added bonus of our project. It has also increased our sadness when we hear about their passing.

In choosing the excerpts for this volume, our goal was to provide a good cross-section of the disciplines that have figured in our interviews, covering physics, chemistry, and biomedicine. The organization of the book is simple: we grouped the entries in three sections: physicists, chemists, and biomedical scientists. Within each section, the entries are arranged in alphabetical order. Many of the scientists could belong to more than one section and some to all three. For a few, the classification may appear a shade arbitrary, but we wanted to restrict the number of sections to three lest the grouping become too fragmented.

In selecting the excerpts, we tried to cover broad ranges of themes. In some cases, we chose scientific topics; in others, personal issues, and so on.[2] There was no grand scheme governing our selections, and in hindsight, we noticed that some topics occur more often than others; for example, reflections on mentors.

We mention here a few topics just to whet the reader's appetite:

- Werner Arber on whether science might destroy life on Earth
- Paul Berg, Benno Müller-Hill, and James D. Watson on the almost taboo subject of the link between genetics and intelligence
- John W. Cornforth about the challenges of his deafness
- Francis H. C. Crick about the right of the incurably ill to terminate their own life
- Paul J. Crutzen and F. Sherwood Rowland about chemistry's influence on the Earth's atmosphere
- Mildred S. Dresselhaus and Christiane Nüsslein-Volhard about women in science
- Freeman Dyson, Vitaly Ginzburg, Paul Nurse, Glenn T. Seaborg, Nikolai Semenov, Gunther S. Stent, and Harold E. Varmus about the science of the future
- Gerard 't Hooft about his disbelief in extraterrestrial intelligence

---

[2] The excerpts from the original published conversations begin and end with quotation marks; the interviewer is quoted in italics and the interviewee in Roman type. Our new comments for this edition are also in italics.

- Isabella Karle about the gap between top science and public ignorance in the United States
- Matthew Meselson on bioterrorism
- Wolfgang K. H. Panofsky on the social responsibility of scientists
- Roger Penrose about the next steps in learning about the world
- John C. Polkinghorne about being a physicist *and* a priest
- Charles H. Townes about the interaction between religion and science
- and much, much more.

Complementary stories sometimes appear in pairs of entries

- Jocelyn Bell Burnell's participation in Antony Hewish's Nobel Prize–winning discovery of pulsars (the Bell and Hewish entries)
- The discoveries leading to the treatment of Parkinson's disease (the Carlsson and Hornykiewicz entries)
- A mentor's generosity facilitating his former associate's Nobel Prize–winning discovery (in the Ketterle and Pritchard entries)
- Two scientists' opposing views of the quasicrystal discovery (the Pauling and Shechtman entries)
- And the bumpy roads traveled by two scientists, independently of each other, on the way to the discovery of magnetic resonance imaging (the Lauterbur and Mansfield entries)

Had we included only a few scientists, we would have inevitably appeared to be biased. However, this is a relatively large set of 111 scientists from sixteen countries (even more if we consider the native countries of immigrant scholars), over one-tenth of them women and over two-thirds Nobel laureates. They cover a large variety of topics and issues and thus provide a good mix.

It gave us great pleasure to reread these conversations and select the excerpts. In this compilation, we again express our respect for our interviewees. We appreciate the advice we received from Dr. Eszter Hargittai and Dr. Michele R. S. Hargittai. We thank our institutions for their support of this project: Saint Francis University in Loretto, Pennsylvania; Budapest University of Technology and Economics; and the Hungarian Academy of Sciences. Special thanks are due to our editor Jeremy Lewis for his trust and encouragement, and to Oxford University Press in New York for bringing out this volume.

> Balazs Hargittai, Magdolna Hargittai, and Istvan Hargittai
> Loretto, Pennsylvania, and Budapest, July 2013

Great Minds

SECTION 1
Physicists

ZHORES I. ALFEROV

In science, my hero was always Academician Ioffe.

Zhores I. Alferov with Mrs. Alferov (on the right) and Magdolna Hargittai in 2001 in Stockholm.
(photo by I. Hargittai)

Zhores I. Alferov[1] (1930–) is Director Emeritus of the Abram F. Ioffe Institute of Physical Technology of the Russian Academy of Sciences, St. Petersburg, Russia. In 2000, he was co-recipient of the Nobel Prize in Physics. He shared one half of the prize with Herbert Kroemer of the University of California at Santa Barbara, "for developing semiconductor heterostructures used in

---

[1] Magdolna Hargittai and Istvan Hargittai, *Candid Science IV: Conversations with Famous Physicists* (London: Imperial College Press, 2004), 602–19.

high-speed- and opto-electronics."[2] The other half of the prize went to Jack S. Kilby of Texas Instruments "for his part in the invention of the integrated circuit."[3] We recorded our conversation in December 2001 in Stockholm, during the week of the Nobel Prize Centennial celebrations.

Alferov made his contributions to the field of communication technologies at Soviet time. The Soviet Union was not considered to be advanced in communication technologies, or if work was being done, it primarily concerned military applications, and all of it was supposed to be classified. Besides, interactions between Soviet and Western scientists were rare. It could have been that Alferov's research had suffered from the isolation, but he did not quite see it that way.

"It may well be that we overestimate this isolation. The isolation was strong indeed right after the war and through the mid-1950s. However, I did my major work in the 1960s, and the isolation, at least in scientific work, was no longer watertight.

International exchange and cooperation has existed in our Institute from the very beginning. The Institute was founded by Academician Abram F. Ioffe in 1918, right after the Revolution. In 1921, he made his first trip abroad in the Soviet time. He received money for this trip directly from Lenin, to buy equipment and literature. When the Institute opened in 1923, it was equipped as well as the best European laboratories at that time. The 1920s and 1930s were a golden age for physics in general, and it was also a golden age for physics in our Institute. All the associates were young. All the physicists who later became famous in the Soviet Union came from Ioffe's School. Nobel Prize winners Landau, Kapitsa, and Semenov came from his Institute. Semenov was Ioffe's deputy for administration. The leaders of our nuclear weapons program came from this school—Kurchatov, Aleksandrov, Khariton, Zeldovich. In our country, the highest award used to be Hero of Socialist Labor. There were only six scientists who got this award three times for the development of nuclear weapons, and five of them were from the Ioffe Institute. In the mid-1930s it was broken, but it was rejuvenated in the 1950s. The traditions for international cooperation were very strong. Even under the Cold War conditions we had good relationships between American and Soviet physicists. The governments on both sides created a lot of barriers, but we frequently succeeded in overcoming them. In the 1960s, the conditions vastly improved. The majority of our journals appeared in English translation, with about a six-month delay. We participated in international conferences, and we invited foreign scientists to our conferences to the Soviet Union.

Wasn't much of the so-called sensitive work classified? The Sputnik, for example, came as a surprise outside the Soviet Union.

There was some secrecy; there was definite secrecy in our work on nuclear weapons and on satellites, and on missiles that would carry the nuclear weapons. But in

---

[2] "The Nobel Prize in Physics 2000," *Nobelprize.org*, Nobel Media AB 2013, http://www.nobelprize.org/nobel_prizes/physics/laureates/2000/.
[3] Ibid.

many institutes of the Academy of Sciences, from the end of the 1950s, beginning of the 1960s, this classified work gradually reduced. If you look at the Russian Nobel Prize winners, the majority were physicists, and all of them came from the Lebedev Institute in Moscow and the Ioffe Institute in Leningrad and, of course, from the Institute of Physical Problems, where Kapitsa used to work. This did not happen by accident because these institutes were complex physical institutions which carried out research in all modern branches of physics, from the beginning. Before the war, they operated openly and had excellent international cooperation. Many of their scientists participated in developing the nuclear weapons program. Some of them left the Lebedev Institute and the Ioffe Institute and became the leaders of the new, secret, classified center for the nuclear weapons program. In the Ioffe Institute and in the Lebedev Institute, classified research was a minor component of the work from the end of the 1950s. When I came to the Ioffe Institute and carried out my work on semiconductor technology, it was classified. Then, from 1955–56, I sometimes participated in classified research for some applications of my former results; but I carried out practically all my research in an open way and was allowed to publish my results. I started participating in international conferences in 1960, and I was not alone. There were plenty of young scientists at the semiconductor physics conference in Prague in 1960. The Soviet delegation consisted of seventy people with the majority being young, around thirty years of age."[4]

We asked Alferov to compare the Soviet and the post-Soviet Russian systems from the point of view of science.

"Today's circumstances are definitely more difficult. The present system is much more bureaucratic than it was before. Undoubtedly, democracy has its merits. You can speak about everything. But it is a strongly bureaucratic system and much of the important things have been captured by the people whom we call the oligarchy. The budget of the whole country has also diminished due to the well-known so-called shadow economy and because of the thieves in the national economy. Science used to have a more important position and a more important role than it has today. This importance was, to a great extent, for the benefit of the military, but not only. I like to note jokingly that in the past, the relative importance of a person or an organization could be determined by the positioning of the names signing obituaries. First came always the members of the Politburo of the communist party and the president of the Soviet Academy of Sciences followed immediately. Only after that came the names of deputy prime ministers and the rest of the crowd. This is no longer the case. The relative importance of science has diminished."[5].

---

[4] *Candid Science IV*, 604–5.
[5] Ibid., 617–18.

PHILIP W. ANDERSON

Science is not just an evolutionary tree; the branches grow back together, they interconnect
—Anderson quoting Steven Weinberg

Philip W. Anderson in 1999 at Princeton University. (photo by I. Hargittai)

Philip W. Anderson[1] (1923–) was born in Indianapolis, Indiana. He earned his degrees at Harvard University, and his career included positions at the US Naval Research Laboratory, Bell Laboratories, and Cambridge University (part-time). He has been at Princeton University since 1975. He has broad interests in physics, especially condensed phase physics. Anderson shared the 1977 Nobel Prize in Physics with Nevill F. Mott and John H. Van Vleck "for their fundamental theoretical investigations of the electronic structure of magnetic and disordered systems."[2] We recorded our conversation in 1999 at Princeton University.

---

[1] M. Hargittai and I. Hargittai, *Candid Science IV*, 586–601.
[2] "The Nobel Prize in Physics 1977," *Nobelprize.org*. Nobel Media AB 2013, http://www.nobelprize.org/nobel_prizes/physics/laureates/1977/.

Anderson was a fierce opponent of President Reagan's Strategic Defense Initiative and was critical of Edward Teller for his contribution to the development of the hydrogen bomb; yet he had no qualms about the American atom bombs deployed to end World War II.

"I'm not one of those who feel guilt about dropping them on Japan. The one thing that emotionally influences me is that I knew about something which most Americans don't, because, having been there, I knew about the fire-bombing from my Japanese friends. The firebombing of Tokyo was so close to genocide, killed so many people, that it seemed to me much more of a horror than the atom bombs. Another thing I was conscious of, and I don't know why so few Americans are conscious of it, is Nanking. Nanking and the Japanese behavior in China and Korea was a horrible thing, unbelievably savage. I don't think I have any complaint whatsoever about the atom bombs. And I'm not sympathetic to the Germans about Dresden. The old saying is absolutely right: 'He that soweth the wind shall reap the whirlwind.' That's what both the Germans and the Japanese did. The bombs left them with no illusions about being defeated."[3]

The following is an extract from our conversation regarding the relationship between religion and science.

"There have been publications, including successful books, linking science and religion. There is also a suggestion that the need for quantum mechanics in describing nature is a manifestation of divine reality. Of course, I do not intend to offend you if you are religious.

I am very much not; in fact, I am an atheist—and I disapprove of the attempt to insert religious meaning into science. In a way, you do offend me a little bit by bringing up religion. I'm an admirer of Weinberg on such questions. On the other hand, I am eager to discuss why quantum mechanics seems so incomprehensible to many people. The problem is in epistemology and semantics. The question is, what is to understand? In doing that, you also must bring in your intuition. Your brain is constructed to see the three-dimensional world with temporal ordering and full of objects. So that is what our brain can comprehend directly. But that is not necessarily the only possible structure that reality can take. That is the only structure that you can comprehend intuitively. Quantum mechanics does not deal with objects. It deals with fields. It is still three-dimensional, thank God, but it might have also had thirty-three dimensions. The minute you realize that distinguishable objects may not be the necessary way to describe how the world works, quantum mechanics is a perfectly deterministic theory of the things it deals with.

Why is there not more open discussion of this?

Partly because even most physicists do not understand quantum mechanics. Broken symmetry is an example where quantum mechanics and everyday reality part company. Quantum mechanics tells us that all eigenstates of systems with a symmetry must be classifiable by that symmetry. But everyday objects, like a pencil, do not represent quantum-mechanical eigenstates. No macroscopic object can be a

---

[3] *Candid Science IV*, 594–95.

quantum-mechanical eigenstate, because macroscopic objects have definite orientation and position in space, and they are a complicated mixture of eigenstates. But they are not moving, they are not time dependent, and thus those eigenstates are enormously degenerate, and that is the phenomenon of broken symmetry. What is really complicated is not the behavior of the microscopic object, say, an electron undergoing interference, but the quantum mechanical description of the macroscopic apparatus with which you measure it."[4]

At the end of our conversation, he responded to a question about the future.

"My wishful thinking is that physics will spread out more toward complexity, geophysics, cosmology and astrophysics, and most of all, biology.

I recently read a book, Ed Wilson's *Consilience*, which impressed me, not so much by teaching me new things, but because it resonated with my thinking. The fundamental concept is that emergence is the way the world works. It is reasonable to study by analysis, to study smaller and smaller pieces of matter, to try to analyze how biology works in terms of the molecules. But there is then another way of looking and thinking of how the complex world arises out of the simpler world. The title of an article I wrote a long time ago is 'More Is Different,' and it is about how, when you put things together, you create more than just their sum. You can follow this through from atoms to molecules and to solids.

Another important thing is that it is all tied together. Ed Wilson calls it 'consilience,' I use the term 'seamless web of science,' which contains all the different fields fused together by intellectual contacts. The fascinating thing is making the connections. Steve Weinberg said recently that science is not just an evolutionary tree; the branches grow back together, they interconnect. If this is true, then physics is not dead because physics is the one science that mixes in with almost everything."[5]

---

[4] Ibid., 597–98.
[5] Ibid., 600.

JOCELYN BELL BURNELL

You can actually do extremely well not getting the Nobel Prize.

Jocelyn Bell Burnell in 2000 at Princeton University. (photo by M. Hargittai)

Jocelyn Bell Burnell[1] (1943–) is one of the most famous Nobel misses. She was born in Belfast, Northern Ireland, and did her doctoral studies with Antony Hewish at Cambridge University (UK). She was instrumental in the discovery of pulsars. In 1974, Hewish received the Nobel Prize "for his decisive role in the discovery of pulsars."[2] There was a general feeling that Jocelyn Bell should have been included. We recorded our conversation in 2000 at Princeton University, where she was a visiting professor in the physics department.

Bell's description of the discovery of pulsars.

"This discovery was an accident because such objects had never been dreamt of. They were unimaginable, literally. I was studying quasars, which are very, very distant

---

[1] M. Hargittai and I. Hargittai, *Candid Science IV*, 638–55.
[2] "The Nobel Prize in Physics 1974," *Nobelprize.org*. Nobel Media AB 2013, http://www.nobelprize.org/nobel_prizes/physics/laureates/1974/.

Bell Burnell with the radio telescope for detecting the signals from pulsars in Cambridge, UK, in the mid-1960s. (courtesy J. Bell Burnell)

objects. The analogy I sometimes use is that you are making a video of a sunset from some vantage point. You have a splendid view of the setting sun. Then along comes a car, and it parks in the foreground and has its hazard warning lights going and thus spoils the picture that you are making. It was a bit like that with us. We were focusing on some of the most distant things in the universe, and this peculiar signal popped up in the foreground. It turned out to be the pulsars, but it was quite a long trail to realizing it. At first, we suspected that there was something wrong with the equipment. Then we suspected that we were picking up artificial interference. We suspected all sorts of things. It was really only when I found the second one that we began to believe that these things might be stars, that they might be natural.[3]

First they were called LGM for Little Green Men.

The naming was in jest, it was a joke, it was tongue in cheek. But radio astronomers are aware that if there are intelligences out there, Little Green Men, then it is they, the radio astronomers, who will probably first pick up the signals. That is the basis behind a lot of the SETI project. But the chances of detecting Little Green Men are very, very, very small, so it was not a serious belief. Of course, when I found the second one and then numbers 3 and 4, the chances totally vanished. It was quite a relief.[4]

What are the pulsars?

---

[3] *Candid Science IV*, 641.
[4] Ibid.

Pulsars have another name, which is 'neutron stars,' because they are made very largely of neutrons. They are very compact and therefore very dense. There is something like a thousand million million million million tons of material, the mass of the Sun basically, all packed into a ball that has a radius of ten kilometers. So their density is comparable to the density of the nucleus of the atom. They are quite extreme and for that reason have some very unusual physics. The ones that we see as pulsars seem to have some very strong magnetic fields. We believe that a beam of radio waves, a bit like a lighthouse beam, is formed at the magnetic poles of the neutron star. The magnetic poles, just as on the Earth, are not coincident with the rotational or geographic poles. On the Earth the north magnetic pole is in Northern Canada. There could be something similar on a neutron star, but maybe even more so with the magnetic poles, say, in Texas. As the star spins, the beam that comes out from the magnetic pole sweeps around the sky, like a lighthouse beam sweeps across the ocean. Every time the beam passes over planet Earth, we can pick up a pulse, so we get a series of regular pulses. Pulsar is an abbreviation for pulsating radio star because of that behavior."[5]

About not sharing the 1974 Nobel Prize.

"The Nobel Prize was awarded thirty years ago. At that time, there was still around the picture that science was done by great men (and they were men). These great men had under them a group of assistants, who were much more lowly and much less intelligent, and were not expected to think. They just carried out the great man's instructions. Maybe that was the way science was done 100 years ago or maybe even more recently. What has happened in the last thirty years is that we've come to understand that science is much more a team effort, with lots of people contributing ideas and suggestions. But at the time of the Nobel Prize, there was still around the idea that science was done by great men, and the awarding of prizes, any prizes, was consistent with that picture. We did not recognize the team nature of science in those days."[6]

Her reaction at the time.

"I was very pleased, mainly for political reasons. I am a strategist, a politician. That was the first time that a Nobel Prize in Physics was awarded to anything astronomical. There is, of course, no Nobel Prize specifically for astronomy, and the physics prize is the nearest. I think it is perhaps debatable whether astronomy is or isn't included in the physics prize. This was the first time that astronomy was included and that was extremely important. It was the opening of a door to a whole new range of things. So I saw that instantly, and I was very, very pleased for that reason.

---

[5] Ibid., 642–43.
[6] Ibid., 652.

Did it not occur to you that you should have been included?

I was content. And, I have discovered subsequently, that you can actually do extremely well not getting the Nobel Prize and have a lot of fun, too. And I have received lots of other prizes."[7]

There is a postscript to the story of Bell Burnell's missing out on the Nobel Prize. Joseph Taylor and his former doctoral student Russel Hulse received the 1993 Nobel Prize in Physics for the discovery of a new type of pulsar. Bell Burnell was invited to attend the ceremony (she had been absent in 1974). At one point, Anders Bárány, the secretary of the physics committee, took her aside and told her that she should have been a co-recipient of the Nobel Prize in 1974. He gave her one of his own small, replica Nobel Prize medals, saying, 'This is the best I can do.'[8]

---

[7] Ibid., 653.
[8] I. Hargittai, *The Road to Stockholm: Nobel Prizes, Science, and Scientists* (Oxford: Oxford University Press, 2002), 240.

CATHERINE BRÉCHIGNAC

After having been chosen I simply had to prove that I could do the job better than anybody else, men and women alike.

Catherine Bréchignac in 2000 in Orsay, France. (photo by M. Hargittai)

Catherine Bréchignac[1] (1946–) was born in Paris. She received her PhD degree in physics from l'Université Paris-Sud in 1977. She has worked at the Laboratoire Aimé Cotton in Orsay since 1971, serving as its director between 1989 and 1995. She was Director General of Le Centre national de la recherche scientifique (CNRS) between 1997 and 2000 and its president between 2006 and 2010. Among her distinctions, she is a member and the permanent secretary of the French Academy of Sciences. We recorded our conversation in her office in Orsay in 2000.

Bréchignac had an unusual reason for choosing physics as her major.

"Originally I was interested in mathematics because I did not have to spend too much time with it—it was logical and easy to do. The other subject I really liked was French

---

[1] M. Hargittai and I. Hargittai, *Candid Science IV*, 570–85.

literature—that was my true love. These were the two things I liked: mathematics and reading.

Later, I decided to orientate toward science, I entered the École Normale in mathematics and finally decided to do physics. It is difficult to say why I did that. I think it was mostly because among the students I knew, I preferred the physicists; they were more social and provided a much more pleasant environment. I made this choice purely based on human interactions and not on the scientific disciplines themselves. Then I decided to try research and went for a PhD in physics. I found research exciting, and by the time I got my PhD, I knew that this is what I wanted to do."[2]

Her main research project was the study of metal clusters.

"Clusters are the component precursors of the nano world. With increasing their size, they can be imagined as bridges between the gas phase and the solid phase. However, because small is different from large even if the properties of solids are usually known, the properties of these clusters are not. So I have decided to work with these clusters by entering a terra incognita, and it has been a real adventure."[3]

During the study of the fragmentation of metal clusters, Bréchignac and her colleagues discovered Coulombic fission in free metal clusters.

"One possibility to induce the fragmentation of a small droplet is to charge it. Since Rayleigh, in 1872, who studied the stability of a drop of water, it was commonly accepted that small multiply charged particles are not stable below a critical size because the Coulombic repulsive energy between the positive charges exceeds the binding energy of the particle. However, there was yet no clear picture concerning the mechanism of fragmentation and the dissociation channel of small clusters when we started. We built a special experiment with two time-of-flight mass spectrometers in a row. The first one is used to select a given size of doubly charged sodium clusters, the second one to study the fragmentation products. We have shown that the critical size, below which multiply charged clusters are not observable in mass spectrometry, strongly depends on the cluster formation. For hot clusters, we established the competition between evaporation of atoms and Coulombic fission in two singly charged fragments. For cold clusters, fission always dominates. The fission products depend on the relative ratio between the Coulombic and the surface energies.

It was not obvious that metal clusters should behave like atomic nuclei, but they do, and the model that describes the fission of unstable nuclei can be successively applied to describe the Coulombic fission of metallic clusters. Moreover, clusters are in some respect more flexible since one can independently vary their charge and their mass that is not the case for nuclei.[4]... The production of materials, devices or systems, by controlling the materials at the nanometer scale, is in full expansion today, and the

[2] Ibid., 571–72.
[3] Ibid., 572.
[4] Ibid., 575.

miniaturization is one of the driving forces of the technology. Research on clusters presents a more conceptual approach. When isolated, the cluster can be considered as the prototype of a small finite system, and an ideal object for scaling laws. When interacting, it offers the possibility to use it as an elementary building block for more complex structures. One of the main advantages of using clusters as building blocks is that they can already be themselves composite systems."[5]

Bréchignac's husband, Philippe Bréchignac, is also a physicist.

"My husband has always helped me a lot and pushed me to succeed. He knows that I am most happy when I work; he knows that I like the research I am doing, and he is very open-minded and understanding. He does not complain if I come home late, if dinner is not ready; on the contrary, he helps to make everything easy. Of course, we have a paid help at home, because both of us are away from home very often. Sometimes he says that too much is too much and then we decide to do something together."

She is a member of the French Academy; her husband is not.

"When I was elected, he was very proud; I could read it in his eyes. But our life together is not based on an academic success. We try to separate our life from our work and avoid having a competition between us. Although, recently, we actually did have some joint work, and we have two papers together. But earlier I tried to avoid that. You should ask him what he thinks about this; but I don't think that it is a real problem for him.

Do you think that your election to the Academy and all your other recognitions were solely on merit or perhaps you being a woman had something to do with it?

Oh, sure, of course. More precisely, about the academic recognition, I don't know; I don't think so. But the appointment to be Director General of the CNRS definitely happened because I am a woman. Of course, I have a good reputation as a scientist, and I also managed the lab and the Department of Physics at CNRS well earlier; but the politicians like to have pride in showing how considerate they are with minorities.

Did you mind that?

No, not at all. I figured that it was fine but after having been chosen, I simply had to prove that I could do the job better than anybody else, men and women alike."[6]

---

[5] Ibid., 576.
[6] Ibid., 577–79.

JOHN H. CONWAY

This mathematical world is consistent.

John H. Conway in 1999 in Auckland, New Zealand. (photo by I. Hargittai)

John H. Conway[1] (1937–) was born in Liverpool, England. He is John von Neumann Professor of Applied and Computational Mathematics at Princeton University and Fellow of the Royal Society (London). He has been active in broad areas of mathematics, and his interests sometimes overlap with those of several of the scientists figuring in this volume because of his contributions to the applications of the symmetry concept. We recorded an extensive conversation with him in 1999 at the University of Auckland in New Zealand, where he and two of us (IH and MH) happened to be visiting professors at the same time.

Conway told us about the nature of mathematical discovery at the end of our conversation, when we asked him whether he had a message he would like to convey.

"There is something. It's how I feel about mathematical discovery. You're wandering up and down, it's like wandering in a strange town with beautiful things. You turn

---

[1] Balazs Hargittai and Istvan Hargittai, *Candid Science V: Conversations with Famous Scientists* (London: Imperial College Press, 2005),16–35.

around this corner and you don't know whether to go left or right. You do something or other, and then, suddenly, you happen to go the right way, and now you are on the Palace steps. You see a beautiful building ahead of you (and you didn't know that the Palace was even there). There's a certain wonderful pleasure you get on discovering a mathematical structure. It happened to me tremendously when I discovered the surreal numbers. I had no idea that I was going to go in there at all. I had no idea of what I was doing. I thought I was studying games, and suddenly I found this tremendous infinite world of numbers. It had a beautiful simple structure, and I was just lost in admiration of it, and in a kind of secondary admiration of myself for having found it. I was so pleased that I'd found it. For about six weeks I just wandered about in a permanent daydream. What happens after that is that I'm vainly trying to re-create that in the people I'm trying to talk to about it, trying to show what this wonderful thing is like and how amazing it is—that you can reach it by studying something else. I'm perennially fascinated by mathematics, by how we can apprehend this amazing world that appears to be there, this mathematical world. How it comes about is not really physical anyway, it's not like these concrete buildings or the trees. No mathematician believes that the mathematical world is invented. We all believe it's discovered. That implies a certain Platonism, implies a feeling that there is an ideal world. I don't really believe that. I don't understand anything. It's a perennial problem to understand what it can be, this mathematical world we're studying. We're studying it for years and years and years, and I have no idea. But it's an amazing fact that I can sit here without any expensive equipment and find a world. It's rich, it's got unexpected properties, you don't know what you're going to find, you might just turn the corner, you might find yourself on the steps of a palace, and you might not.

I can't comprehend how this can be. I don't know what it means. I don't know whether there is such an abstract world, and I tend not to believe there is and to believe that we are fooling ourselves.

We used to think that the earth is flat, and it was inconceivable that it could be round. It was only some very painful facts that eventually forced us to believe that the earth is roughly spherical. What's happened continually in the physical sciences is that the truth was not one of the possibilities that was considered and then rejected, not even that. It was one of the possibilities that couldn't even be considered because it was so obviously impossible.

In mathematics our development has come a little bit later, but the same sort of thing happened with Gödel's theorem, and so on. What we thought was the truth was just a kind of approximation to the truth. Newtonian dynamics is an approximation to relativistic dynamics, and it's not literally true if you go to high speeds and high energies; if you go to small distances, it doesn't work quite either, according to the quantum theory. In mathematics we have these beliefs that there are infinitely many integers and so on. Any belief like that about the nature of things that are arbitrarily far away has turned out to be false in physics. So I think it is false in mathematics, too. I think that eventually we'll find something wrong with the integers, and then the classical integers will be just an approximation. That's a big puzzle for me. I don't quite believe in this artificial mathematical world. There appears to be a wonderful consistency about it: I can think of something in some way; someone else can think about it in a different way; and we both come to the same conclusion. If we don't, there must be a mistake—at least it has been so, so far. But I don't see why there should be this consistency in a world that I don't really believe exists. So to me it's a sort of fairy tale and fairy tales don't have to be consistent because they are human creations ultimately. But this mathematical world is consistent and I wonder, 'What the hell is it?' without implying the works of anything supernatural. I'm nonreligious."[2]

---

[2] Ibid., 33–35.

MILDRED S. DRESSELHAUS

Women have a lot of bad things that happen, but they just have to work around them.

Mildred S. Dresselhaus in 2002 at MIT. (photo by M. Hargittai)

Mildred S. Dresselhaus,[1] (born Mildred Spiewak in 1930 in Brooklyn) is Professor of Physics and Electrical Engineering Emerita and Institute Professor at the Massachusetts Institute of Technology. She studied at Hunter College in New York City and received her master's degree in physics from Harvard University. She did her doctoral work at the University of Chicago and completed a postdoctoral stint at Cornell University together with her husband, fellow physicist Eugene Dresselhaus. Mildred Dresselhaus has researched materials for her entire scientific career and has always been at the forefront when new discoveries in materials science were made, especially in carbon science. Mildred and Eugene have four children.

Mildred Dresselhaus has been very active in professional societal life and is one of the most decorated scientists in the United States. Her memberships

---

[1] M. Hargittai and I. Hargittai, *Candid Science IV*, 546–69.

Mildred S. Dresselhaus around1960 at Cornell University. (courtesy of M. S. Dresselhaus)

include the National Academy of Sciences; and her awards the National Medal of Science. We recorded our conversation in her office at MIT in 2002.

Early on in her career, in 1958, Dresselhaus taught a course on electromagnetic theory to engineering students at Cornell University. A woman teaching such a class was unusual at the time.

"The reason I taught this course was that the professor who was supposed to teach the course left during the first week of the semester and they couldn't find a person to teach that course. I volunteered to teach the course, with no pay, because I had a fellowship. There was a big uproar. The faculty met every day for a whole week to decide, not whether I was qualified to teach the course, but whether the young men would pay attention to me as a young woman. I had a lot of experience in the area of electromagnetic theory. There were no women in the course. It was difficult for these senior faculty members to comprehend and deal with having a young woman teach young men. Maybe they decided that it was OK because I was married and already had a baby. I never knew exactly what went on behind the closed doors. However, I did get a chance to teach the course, and I did a very good job with that course. I found out more about that many years later when various students who were in the course came to me, and they remembered my teaching, because it was somehow different for them. They told me years later how much that course meant to them.

There is then an anecdote from a later time. I was the treasurer of the National Academy of the United States, which is quite an important position in Washington. We were having a meeting of the governing board of the National Academy in the lecture hall. A presenter came from the National Aeronautics and Atmospheric Sciences, one of our US agencies. He walked in; he recognized me after all that time; and he said that before giving his comments, he would like to say something that he had meant to say for many years about the influence I had on him. It was a fantastic introduction and terribly embarrassing for me. People do appreciate a teacher that goes out of the way to do something beyond the call of duty.

At MIT, I was very well treated over the years. Within five years after I joined the faculty, I was appointed Associate Department Head of Electrical Engineering. We had sixty-six faculty members in my department, all of them men. I had a lot of opportunity for leadership and I never felt that people made a big distinction here about my being a woman. That's one good thing about science, a good thing about physics, a good thing about engineering, that there is some kind of standard—comparisons in performance are more objective than in other fields. I have noticed that physics is much more of a macho thing than chemistry, at least in the US; but still, once you're doing the work, it is all right to be a woman. Getting through the door is what is hard, and it's still hard. Just look around, we still have very few women PhDs in physics, about 15 percent. We should be much further along, considering what's happened in all the other fields. In the United States, women in general get PhDs at almost the same rate as men, once they are in physics. So it is not that women would not go on for advanced degrees after college. But they don't choose certain fields. There is still something about certain fields that is not attractive and it's not only the subject matter. I think it's also the sociology of the field. Just across the border from physics, in materials science and in chemistry, proportionally, there are more than twice as many women."[2]

She has been very active in women's issues.

"I can encapsulate how it all started. When I got my first appointment here at MIT, it was to a chair that Abby Rockefeller Mauzé, the sister of the five Rockefeller brothers, established. Her idea must have been to provide an opportunity for women to be scholars. I felt I had to set aside a little time for mentoring women students, and I spent about an hour or an hour and a half per week for counseling women students, trying to help them along. I was doing this from the very outset because part of my appointment, or my interpretation of my appointment, included this kind of activity. However, it has never been a main thing I do here at MIT. Women had very little academic opportunity at that time in the 1960s. Our women students needed role models, so I tried to help.

Rosalyn Yalow was a very good role model for me. Early on, I had one male student, who was very interested in public affairs, science policy, and who was a very good graduate student. He had a girlfriend; she had social troubles at MIT, and her

---

[2] Ibid., 551–53.

academic troubles came from her social troubles. She needed a role model; she needed somebody to help her. He told me that where I was different from the other MIT professors was that I could do something special for the women students here, especially those who needed help. He convinced me that if I were to spend, say, 5 percent of my time, which is one or two hours a week, I could have big impact. I thought about that and I decided that he was right. As a result, I have worked on many projects to help improve the quality of life for women students. There has been a big payoff because they tell me, ten years later, that I'm the reason that they stayed on to complete their degree program. I could probably be helpful because of my own experience. Later, the president of MIT, Jerry Wiesner, during his term, asked me to help women at MIT to succeed, and I did that. Then I was asked to do different things for women at the national level too. I didn't go out of my way looking for this kind of work, but I don't say no to it either, just like I do a lot of public service that has nothing to do with women in science."[3]

---

[3] Ibid., 558–59.

FREEMAN J. DYSON

I would like to be remembered more as a writer than a scientist.

Freeman J. Dyson in 2000 at the Institute for Advanced Study in Princeton, New Jersey.
(photo by M. Hargittai)

Freeman J. Dyson[1] (1923–) is a British-American scientist and author. He was born in Crowthorne, Berkshire, England. He received a bachelor of arts degree in mathematics at Cambridge University in England and then taught physics at Cornell University. He has been at the Institute for Advanced Study (IAS) in Princeton, New Jersey, since 1953. He has been involved in a great variety of projects in physics and has been recognized by many awards and distinctions. We recorded our conversation in 2000 in Dyson's office at the IAS.

---

[1] M. Hargittai and I. Hargittai, *Candid Science IV*, 440–77.

Dyson talked about his career path.

"I had two careers, one as a scientist and one as a writer, and being a writer is just as exciting as being a scientist. In particular, if you are fifty years old, it is much easier. So at the age of fifty, I made the decision more or less to switch. It is clear to me that over the age of fifty it is much harder to compete as a scientist, but it is quite easy as a writer."[2]

Dyson has advocated that physicists should help molecular biology. He prognosticated what the field would look like in thirty years.

"I would think that probably the single molecule sequencing will turn out to be practical and probably will be done mostly by physicists. Of course, you never can tell how long these things will take. I would think that it ought to take less than thirty years, certainly, so we have very fast sequencing where you do not have to separate the molecules. Of course, similar problems are in protein chemistry; there we still do not have a really rapid way of solving protein structures."[3]

On the dangers of combining computer science and genetic engineering.

"Everything connected with biology is dangerous. It is much more dangerous than physics, but nevertheless, I think that it is extremely promising as well. So one has to watch out for the dangers; but it would be stupid not to pursue the promise. The promise is, of course, mostly in medicine, but no doubt it will also be applied to the breeding of animals and plants. This will certainly become even more of an art than it is today, and it could be a really exciting art form to design plants and animals. If you think how much effort goes into growing different kinds of flowers, such as orchids and roses, this could be done better with a certain amount of computer assistance.
It is becoming more dangerous when moving toward the animal world.
Exactly, that is right. The closer we get to humans the more dangerous it is. That is quite right.... Fertility clinics are, of course, the places where you produce babies, and that is the place where the action is. It is now all over the world, not only in the rich counties but in poor countries as well, this is a fast growing branch of medicine. Giving people the chance to have babies is an enormous thing.... It is very quickly going to be possible to manipulate the embryos, to put in genes that you want or take out genes that you don't want. And that is when it really begins to be a problem.[4]
If physics dominated the first half of the twentieth century and biology the second half, what would dominate the twenty-first century?
I usually say neurology, but, of course, I could be quite wrong. The real dominant science may be something that has not yet begun. So we don't know. From among the

---

[2] Ibid., 474.
[3] Ibid., 442–43.
[4] Ibid., 443.

present existing sciences I think it is neurology that is the most likely because it deals directly with the brain and the brain is the big, unsolved problem."[5]

On Carleton Gajdusek.

"First of all, he is a great scientist. He did this very important work on kuru, and also another very interesting disease in Siberia that he is still studying (Viliuisk encephalomyelitis). He was planning to go back there to study it. He used to say that if all else failed he could still go to Siberia and take a job there. As a scientist, he is certainly great. He is also a wonderfully unconventional person. I always enjoy heretics. He adopted sixty Melanesian kids. Four of his adopted kids were there when we greeted him at the jail. They honor and respect him. The fact is that he got persecuted only because this country has an affinity for witch-hunts. They call somebody a child abuser, it is like calling somebody a communist. And then nobody will come to their help, and things get blown up terribly. Gajdusek got into this witch-hunt. Whatever he may or may not have been doing, it was clearly not bad for the kids; they came out of it extremely well. He adopted them, paid for their education; so everything he did was certainly for their good. He has a tremendous sense of humor about this business. When he was in jail he started to write letters, and I got lots of letters from him. They were always full of jokes.

This whole witch-hunt started when two of his kids brought charges against him for sexual abuse. These kids had their own sexual traditions that were very different from ours. Carleton said that if anybody corrupted anybody, it was they [who were] corrupting him and not the other way around. Undoubtedly, they were very free and easy with sex, so according to the strict interpretation of the laws, you can call that child abuse; but the case should never have been brought to court. In fact, what probably happened was that two of the adopted kids who did not do as well as the others, decided that this was the way to blackmail Gajdusek— and they hoped to get rich by blackmail. Anyway, it is a sad story. But on the other hand, it is not so sad because he managed to handle it so well. When he came out of the jail the first thing he said was 'they broke Oscar Wilde but they did not break me.' So he is definitely a hero. I also wrote a piece about him, called 'A Hero of Our Time.' So altogether, I think that Feynman, [Martin Luther] King, and Gajdusek would be a good set of heroes."[6]

---

[5] Ibid., 444.
[6] Ibid., 457–58.

JEROME I. FRIEDMAN

At the end of my junior year in high school... I picked up a book by Albert Einstein entitled Relativity, *and I was absolutely fascinated by it.*

Jerome I. Friedman in 2002 at MIT. (photo by I. Hargittai)

Jerome I. Friedman[1] (1930–) was born in Chicago. He is Professor of Physics Emeritus, and Institute Professor at the Massachusetts Institute of Technology. He was one of the three recipients of the 1990 Nobel Prize in Physics, together with Henry W. Kendall, also of MIT, and Richard E. Taylor of Stanford University, "for their pioneering investigations concerning deep inelastic scattering of electrons on protons and bound neutrons, which have been of essential importance for the development of the quark model in particle physics."[2]

Friedman received all his degrees in physics from the University of Chicago, and spent periods of time there and at Stanford University before joining MIT in 1960. We recorded our conversation in 2005, in his office at MIT.

---

[1] M. Hargittai and I. Hargittai, *Candid Science IV*, 64–79.
[2] "The Nobel Prize in Physics 1990," *Nobelprize*.org. Nobel Media AB 2013, http://www.nobelprize.org/nobel_prizes/physics/laureates/1990/.

Enrico Fermi. (courtesy of Oak Ridge National Laboratory)

Friedman was a student at the University of Chicago when its physics department had a concentration of stellar professors. We asked him about his experience with Enrico Fermi and others.

"He [Fermi] gave a course in the spring of 1954. It was on quantum mechanics and I sat in on the course. When Fermi gave a course, I always sat in on it regardless of whether I had taken it before or not, because I could always learn something from this great man. Fermi was a robust man who appeared to be in excellent health. He went away that summer to Italy and developed a rapidly growing form of stomach cancer. I saw him when he came back in September. He was about fifty feet away from me in the corridor as he was going into his office. I waved to him and he waved back to me. I looked at him and was terribly startled to see how gaunt he looked. The next day he went to Billings Hospital, where he had exploratory surgery and was found to have inoperable cancer. I never saw him again. He died on November 28. What a loss it was. Besides being a great physicist, he was a very kind and considerate man; and he had enormous patience in explaining physics to his students. He was a wonderful human being.

For my thesis research, Fermi had suggested that I carry out a nuclear emulsion investigation of proton polarization produced by nuclear scattering, an effect that had been observed at cyclotron energies. The objective of this study was to determine whether the polarization resulted from elastic or inelastic scattering. I did not know at the time that Fermi had already theoretically shown that elastic nuclear scattering could produce large polarizations. This calculation was in his famous notebook of problems that he had investigated and solved. The calculation was, as usual, based on a simple model, utilizing a real and an imaginary nuclear potential and a spin-orbit coupling term. This is the same term that he had suggested to Maria [Goeppert] Mayer as possibly playing a role in the structure of the nucleus and which was crucial to her development of the Shell Model.

When I had only partially completed scanning my emulsion plates, Segrè visited Fermi and told him that he had observed large polarizations in nuclear elastic scattering in a counterexperiment at the Berkeley cyclotron. According to Segrè, on the morning of his visit, Fermi calculated the polarization expected in this experiment, and his results matched Segrè's measurements beautifully.

I had been scooped and was quite dejected. However, Fermi was very understanding and suggested that I continue my measurements. First, it would be valuable to confirm Segrè's results with another technique; and secondly, I could also determine to what extent inelastic scattering produced polarization.

When Fermi died, I was devastated. What an immense loss it was to all of us. My thesis work was not yet completed, and I clearly didn't want to start all over again on another problem with another professor. John Marshall kindly came to my rescue. He took over my supervision, and he ultimately signed my dissertation.

After I received my PhD, I continued working in Fermi's nuclear emulsion laboratory, which had been taken over by Val Telegdi, who was an outstanding young faculty member. At about that time, there was the so-called Tau-Theta paradox in which there appeared to be two particles of opposite parity having the same mass. These puzzling results were causing much controversy and speculation in the particle physics community. In a bold paper, Lee and Yang proposed that this paradox was due to the non-conservation of parity in the weak interactions and suggested some experimental tests of this hypothesis,

While most of the community considered the conservation of parity to be sacrosanct, Val was quick to pick up the significance of this paper. He asked me to join him in making a measurement of muon decay in nuclear emulsion to test this radical new idea. Following the suggestion of Lee and Yang, we planned to measure a forward-backward asymmetry in muon decay, which would be sure sign of the violation of parity conservation. In my thesis work, I had developed some expertise on how to make double-blind visual measurements of asymmetry. Most of the others in our lab thought that this was a waste of time. I remember giving an Institute seminar on the measurement we were going to make. After the seminar, a distinguished older member of the faculty came up to me and said that I had given a nice talk, but that I should realize that we were not going to find anything.

As it turned out, we were one of the first three groups that demonstrated the non-conservation of parity in the weak interactions. Madame Wu and collaborators were the first to demonstrate this effect in their measurement of the decay of cobalt 60; and Garwin, Lederman, and Weinrich also did so in their measurement of muon decay. These two beautiful counter experiments demonstrated the effect with excellent statistics. While many perceived this as an experimental race, it really wasn't so from our perspective. When we started our experiment in the late summer of 1956, we knew of no other measurements going on at that time to test the non-conservation of parity. Our progress was hampered by the following circumstances. Val had to go to Europe during the early autumn of 1956 on personal matters and remained there for about two months. During this period, I was starting to see a hint of an effect and I wanted to get more scanning help. But the physicist left in charge of the emulsion lab wouldn't give them to me, because the only scanners available were involved in what was thought to be a more promising measurement. Only when Val returned did I get more help. We heard about the other two experiments when our scanning was close to being complete and we were already seeing an effect. All three publications appeared within a short time of one another,

In addition to being a physicist of deep insight and strong opinions, Val has a wonderful sense of humor and an inexhaustible supply of jokes. He was an excellent mentor, and I learned a great deal from him. He also helped me get my first real job, a three-year post-doc position with Robert Hofstadter at Stanford University."[3]

---

[3] *Candid Science IV,* 72–74.

RICHARD L. GARWIN

We are lucky not to already have had a terrorist nuclear explosion in one of our cities. I confidently believe that we will have one within the next few years.

Richard L. Garwin in 1957 with an experiment. (courtesy of R. L. Garwin)

Richard L. Garwin[1] (1928–) was born in Cleveland, Ohio. He did his undergraduate studies at the California Institute of Technology and received his PhD degree from the University of Chicago in 1949. He served in leading positions of the research institutions of IBM Corporation. Simultaneously, he held consulting, advisory, and part-time appointments in defense-related and national

---

[1] I. Hargittai and M. Hargittai, *Candid Science VI: More Conversations with Famous Scientists* (London, Imperial College Press, 2006), 480–517.

Richard and Lois Garwin in 2004 in in Scarsdale, New York. (photo by I. Hargittai)

political organizations, including the Los Alamos National Laboratory. We recorded our conversation in 2004, in the Garwins' home in Scarsdale, New York.

Perhaps Richard Garwin's most remarkable single accomplishment was his design of the "Mike" test of November 1, 1952, in which the Teller-Ulam model of radiation implosion was employed in a thermonuclear explosion. In March 1951, the Teller-Ulam principle was described in a secret report titled "On Heterocatalytic Detonations 1. Hydrodynamic Lenses and Radiation Mirrors." Only the title of this report has been unclassified. The rest of the paper remains classified.

"In February 1951, Ulam, whom I knew well—he was a very gregarious person—had to talk about his work, and he had had an idea. His idea was not necessarily for the hydrogen bomb; his idea was to have an auxiliary bomb that would prepare a main bomb. If it were a matter of ordinary fission explosion, the auxiliary bomb could compress a large amount of material much better than high explosive could, and if it were a thermonuclear fuel, it might help there too. It was an ill-formed idea; and the idea was to use a shock wave from a nuclear explosion to do this. He went to see Edward Teller and Teller said, 'It won't work, and I have a theorem why it won't work in the case of the hydrogen bomb—because everything is a bimolecular reaction in the case of the hydrogen bomb.'

Teller dictated a twenty-page sort-of testament in 1979. He said that if it won't work in the normal density of liquid deuterium, then it won't work in a thousand-times-compressed deuterium either, because all of these rates go up by a factor of million by unit volume and by a factor of thousand by particle number. And Teller said to Ulam, 'By the way, if this would work, there is a better way to do it; we could use radiation, the soft X-rays.' The soft X-rays constitute most of the energy in nuclear explosion to do the same thing. That's the origin of this, really, two-part paper—hydrodynamic lenses to use the shock to squeeze in a very well-defined way, and the radiation mirrors, 'radiation case' as it turned out to be called in the actual implementation. That was on February 24. And Ulam wrote to John von Neumann, who was a consultant to Los Alamos; they were good friends. Ulam said that he [Ulam] had a good idea but it might not work 'because Edward likes it too.' But Ulam did not have any idea of the radiation implosion in that letter, that is, the use of thermal radiation from a nuclear explosion to prepare the main charge.

By March 9, 1951, everything had been worked out by Teller because Stan Ulam was not such a practical, calculating type person as to be able to carry things through. At Los Alamos, people now realized that contrary to what they had thought before, there was a way to build hydrogen bombs. It was not the fact that you use the radiation or you use the shock wave, it was that compression really helped. As Teller says in his testament, that's why we call it the Equilibrium Super; before it had been called just the Super. The name Equilibrium Super implies that there is only so much energy that can be held in the radiation field. In spite of the fact that all the rates go up with compression, the amount of energy that can be held in the radiation field is determined only by temperature. The amount of energy that can be produced by the nuclear materials is proportional to compression. That's why people realized suddenly that was the way to do this. It's not at all clear why it took so long to understand this. Hans Bethe in his later writings called it a miracle of understanding. In fact, Hans or I, or anybody else could easily have made this step, especially since we had a major fusion-related experiment designed by March 1951, and the nuclear explosion tested in the Pacific in the summer of 1951. It was the so-called George test in the Greenhouse series of nuclear tests. George was the particular experiment, but it was not an experiment of the Ulam-Teller idea.

When I went to Los Alamos in May 1951 and asked Teller what had happened in the interim, he showed me this paper that he and Stan Ulam had authored. He then asked me to devise an experiment (presumably a small nuclear test) to prove that it would work or would not work. I decided that the most convincing test would be to actually make a hydrogen bomb using these ideas. I began my work on May 1 and the test was fired on November 1, 1952. It was code-named Mike. It had eleven megatons of energy release; almost a thousand times the thirteen kilotons of energy produced by the Hiroshima bomb and about five hundred times the twenty kilotons of the Nagasaki plutonium bomb.

I did not invent the idea of radiation implosion; I did not invent anything that was in the Mike shot, but I did decide among all of the competing ideas as to how should you fashion this. Whether to use liquid deuterium or a compound of deuterium, lithium deuteride, which is used in all modern hydrogen bombs, this question remained open until January 1952. In Mike, we used liquid deuterium; and the reason was that I felt comfortable with liquid hydrogen, which had been related to my work in particle physics experiments at University of Chicago, and I had no trouble designing this thing (which weighed 70 tons) to have a couple of cubic meters of liquid deuterium, kept cold by liquid hydrogen. The device had an ordinary nuclear explosive at one end, which had to be kept warm because you cannot detonate an explosive when it is cold and there were all kinds of problems like that."[2]

[2] Ibid., 489–91.

VITALY L. GINZBURG

It is the luck of humankind that Stalin and Hitler did not possess atomic bombs first.

Vitaly L. Ginzburg and Istvan Hargittai in 2004 at the Physical Institute of the Russian Academy of Sciences in Moscow. (unknown photographer)

Vitaly L. Ginzburg[1] (1916–2009) was born in Moscow and spent most of his professional life at the P. N. Lebedev Physical Institute of the Russian (earlier, Soviet) Academy of Sciences, Moscow, for some time as the head of the I. E. Tamm Department of Theoretical Physics. Igor Tamm was his mentor, but he considered Lev Landau also as his teacher. He was a full member of the Russian

---

[1] I. Hargittai and M. Hargittai, *Candid Science VI*, 808–37.

Ginzburg around 1950, working on an experiment in optics. (courtesy of Victoria Dorman, Ginzburg's granddaughter)

Academy of Sciences, and in 2003, he shared the Nobel Prize in Physics with Alexei A. Abrikosov and Anthony J. Leggett "for pioneering contributions to the theory of superconductors and superfluids."[2] We recorded our conversation in 2004 in Ginzburg's office at the Lebedev Institute.

On politics.

"There was a famous singer in Russia by the name of A. Galich, who then emigrated; his original surname was Ginzburg, but no relation. His words were something like, 'more than anybody else, be afraid of the man who will tell you how things should be done.' Great words and he meant that Lenin, Stalin, and Hitler preached that they possessed the only right answers to everything. I do know that I don't, but I have my opinion, and here it is: Totalitarianism is worse than anything else. The only acceptable way to go is democracy. Also, I know that in Russia there are liberties that were nonexistent in Soviet times. I agree with the great Churchill that democracy has many problems, but it is still the best form of government. I disagree with those who can

---

[2] "The Nobel Prize in Physics 2003," *Nobelprize.org*. Nobel Media AB 2013, http://www.nobelprize.org/nobel_prizes/physics/laureates/2003/.

complain only and sing the praise of the old regime. There used to be censorship and there is no censorship today, well, there is only partial censorship today, especially for television. There was no freedom to travel; today, there is freedom to travel. Although I am an atheist, I appreciate today the freedom of worship."[3]

On religious instruction in schools.

"I have been an atheist all my life and have not been interested in such debates. Recently I got involved in such a debate by accident. There was an article in our literature magazine, Literaturnaya Gazeta, declaring that there are hardly any atheists left in Russia, more or less suggesting placing all remaining atheists in quarantine. My friend, Evgenii Feinberg and I were so upset that we responded to this article rejecting the suggestion in no uncertain terms. From this, everything then started and I have written copiously about this topic ever since. At the same time I condemn in the strongest terms the crimes the Bolsheviks committed against those serving various churches or simply being religious.

At this time, the suggestion is to offer religious instructions as elective subject in school. However, what does 'elective' mean? It means to take it or not to take it; I find it unacceptable and I am determined to fight it. My wife tries to constrain me; she tells me that at my age—I am 88 years old—this should not be my concern."[4]

Ginzburg listed the next directions in physics several times during the last decades of his life. In this connection the question arose about his personal choice of research if he were twenty-five years-old again.

...I have worked in many areas and especially in theoretical physics. I often did something one day and something very different the next day. If starting anew today, I'm sure I would again become a theoretical physicist. I have a prayer and although I usually do not explain it to people, I am going to explain it to you. As you know, Jewish men have such a prayer in which they thank God that he did not make them into women. In my prayer, I am thanking God for having made me into a theoretical physicist. This does not mean that I have anything against experimental physicists. In my eyes they have the most difficult job possible. They have to sit at some apparatus all their lives. What the great luck of theoretical physicists is that he can easily change his topics all the time.

In my Nobel lecture, I raised this question, why it took me so long to generalize the London equation. I understood that it needed generalization as early as 1943, and yet we published this generalization only in 1950. The reason was that I was busy with many other things. I always dealt with many problems. This is why I cannot give you a specific response to your question about what would be my choice today if I were twenty-five years old again. Problems in theoretical physics would be sufficient to

---

[3] *Candid Science VI.*, 829.
[4] Ibid., 829–30.

keep busy a thousand Ginzburgs. If then taking a closer look at theoretical physics, I do have some fixed ideas. From 1964, I have been interested in high-temperature superconductivity. Today, the question is about room-temperature superconductivity. Could we make a superconductor that it would be possible to utilize at room temperature for which, for example, water-cooling would suffice? This is what I find to be a most interesting problem. It may not be the greatest challenge in physics today, but I would probably select this one to pursue it further if I could suddenly become young again."[5]

[5] Ibid., 830–32.

DONALD A. GLASER

I've been regarded as crazy, many times.

Donald A. Glaser in 2004 at Berkeley. (photo by I. Hargittai)

Donald A. Glaser[1] (1926–2013) was born in Cleveland, Ohio. He was an experimental physicist turned experimental biologist turned neuroscientist and computational biologist. He received his bachelor of science degree in physics and mathematics from the Case Institute of Technology in 1946 and his PhD degree in 1950 from the California Institute of Technology, studying high-energy cosmic rays and mesons at sea level. He taught at the University of Michigan, and in 1959 moved to the University of California, Berkeley, where he remained for the rest of his career. He was both Professor of Physics and Professor of Molecular Biology. In 1960, he was awarded the Nobel Prize in Physics "for the invention of the bubble chamber."[2] We recorded our conversation in the Glasers' home near Berkeley in 2004.

---

[1] I. Hargittai and M. Hargittai, *Candid Science VI*, 518–53.
[2] "The Nobel Prize in Physics 1960," *Nobelprize*.org. Nobel Media AB 2013, http://www.nobelprize.org/nobel_prizes/physics/laureates/1960/.

Donald A. Glaser with Magdolna Hargittai in 2004 in his home near Berkeley, California.
(photo by I. Hargittai)

Glaser's interests covered broad areas of physics and biology.

"It goes further back than that. I began as a musician and then, for about two months, I was an engineering student because neither my parents, nor my teachers knew the difference between engineering and physics. It took me two months to realize that physics, as a profession, was different from engineering. Then I worked seriously in physics. I've been thinking about what I should tell you to answer a question like that. The answer is that there is a thread through my life and it is escaping from big science. I like to work with a group of students and postdocs, but I don't like to be in very large research teams. I began with a thesis in cosmic ray physics, which is a very solitary thing; I did my own little project. Then I was offered a job—when I was first looking for a position—to work with cyclotrons at Columbia or MIT or at some place else where there would've been a large group. Instead, I took a position at Michigan, which is a good university, where I was promised that I could do whatever I wanted to do. I decided that what I really wanted to do was to find out more about the so-called elementary particles of physics.

At that time, the only really productive method was the large cloud chambers. Sometimes high in mountains and sometimes at sea level you just waited hour after hour, day after day, to get by chance a picture which showed some novelty that could not be explained by the current understanding of the particles. Sure enough, there were things, which began to be called *V* particles because they made tracks of the shape *V*. Then the theorists decided that these particles were illegal according to the current understanding and so they began to call them strange particles because they did not obey the theoretical structures. The quest was to look for more of these particles like pieces in a jigsaw puzzle that might make enough to give a global theory of what these particles were and what their role was in describing the dynamics and the structure of the Universe. But it was painstakingly slow and I decided that I would try to device a method for increasing the rate of getting information about this family of particles that we were beginning to discover and by 'we' I mean the world collectively.

My fantasy was that if I was clever enough, and had a good enough instrument, I could sit by myself in a little cabin on the top of a mountain and gradually collect more information. In those days, one could get one interesting picture a day perhaps. Then with a slide rule, we could calculate a little bit of relativistic dynamics that you had to do to measure the mass and charge of the particle. We didn't use computers, and it wasn't necessary. Later, when computers were available, of course, that made it a lot easier. I went through a series of inventions, which were particle detectors of various kinds and finally, the bubble chamber, which increased the rate of collecting information by about a thousand fold. But a very disappointing thing was that it was no good to use on top of a mountain but was ideally suited to use at big accelerators. So I was trapped.

In order to advance the subject that I was interested in, I was forced to work at the big accelerators. I did this for a number of years, but they got worse and worse, and each experiment in those days cost twenty or thirty million dollars, so you could not have an impulse and go to the lab in the morning and tell your students about an idea and do something about it. Instead, it was committee after committee and weighty decisions and so on. Finally, we did generate large numbers of pictures, which contained interesting events. They were so many that we couldn't analyze them and it became a question of automating the pattern recognition and diagnosis software and hardware. We built scanners and software and studied the question of how can humans recognize patterns so readily and computers can't. In those days, pattern recognition was very unsuccessful. I sent pictures all over the world and a lot of universities were looking at our pictures and finally we had to meet in Geneva in order to agree on a final draft of a paper by 23 authors. At that point, I decided that that was it."[3]

---

[3] *Candid Science VI*, 520–22.

MAURICE GOLDHABER

When I became a physicist, science was a calling; later, for many, it just became a profession.

Maurice Goldhaber in 2001 at Brookhaven National Laboratory. (photo by I. Hargittai)

Maurice Goldhaber[1] (1911–2011) was an Austrian-born American physicist. He studied at the University of Berlin and in 1933 he became a refugee scientist first in England, then in the United States. He received his PhD degree in the Cavendish Laboratory in Cambridge, UK. He started his career at the University of Illinois at Urbana, and was at the Brookhaven National Laboratory from 1950 to the end of his life. He is best known for his research achievements in nuclear physics and the physics of fundamental particles. In 1939 he married Gertrude Scharff, subsequently a most distinguished physicist in her own right. We recorded our conversation at Goldhaber's office at the BNL in 2001. In 2002, we returned for a second visit to augment the conversation.

---

[1] M. Hargittai and I. Hargittai, *Candid Science IV*, 214–31.

Gertrude and Maurice Goldhaber. (Brookhaven National Laboratory, photography division. Courtesy of M. Goldhaber)

Maurice Goldhaber was a member of the great generation of physicists who could still enjoy the sizzling atmosphere of the Berlin colloquia, and then he moved to the United States to participate in building up a world-class community of physicists. This is how he narrated the beginning of his career.

"I was born in Lemberg, which at that time was in Austria-Hungary. Then it was successively in Poland, the Soviet Union, and now [as Lviv] in the Ukraine. My parents spoke Polish, Yiddish, and German. At the end, my father spoke fourteen languages. When I was three, we left for Egypt, where my father became a travel agent. When my parents did not want us children to understand them, they spoke Italian. So when Fermi's papers appeared written in Italian, I could read them. During World War I the British interned my father as an enemy civilian. My mother and us children were sent to Austria by boat. My maternal grandparents lived then not far from Vienna in what later became Czechoslovakia. There I started school. The instruction was in German. My father returned after the war and we moved to Chemnitz in Germany. I graduated from 'high school,' the Real gymnasium, in 1930. Then I went to the University of Berlin to study physics. In the colloquium there were Max Planck, Albert Einstein, Max von Laue, Walter Nernst, Erwin Schrödinger, Otto Hahn, and Lise Meitner in the first row.

There were also, often, famous visitors. Among the younger staff and the students there were also many future famous scientists, including Leo Szilard. It was very stimulating. In 1933, everything changed with the Nazis coming to power. My father and the rest of the family left for Egypt, but I stayed in Berlin. I didn't want to leave until I was accepted somewhere else. Ernest Rutherford accepted me as a research student in the Cavendish Laboratory in Cambridge; so in June 1933 I arrived in England. First, I stayed in London and went to Cambridge in August. I was the first refugee in Rutherford's laboratory. Later on others came. Rutherford was active in the movement of saving refugee scientists. Szilard was instrumental in starting the Academic Assistance Council. Back in Berlin, Szilard was one of the few older people who would talk to the young students. We met in London occasionally and discussed world affairs and nuclear physics. In time we developed a 'mutual admiration society'. At one time, when we worked on a nuclear problem while Szilard was in Oxford, a British physicist, passing by asked, 'who of you is the brighter?' I was struck dumb, but Szilard deflected the question and pointing to me, said 'He has more imagination.' Thinking of this remark I realized that 'my strength is my weakness, my weakness is my strength.' When I read the Szilard biography, *Genius in the Shadows*, I was surprised to learn that he had converted to Calvinism at age twenty-one. I knew him well enough to know that it was not 'internal conversion.'

In Cambridge, I met James Chadwick who looked after the students, especially when Rutherford was away—as happened on my first visit. Chadwick was a man of few words. I mentioned to him that I would like to join the inexpensive Fitzwilliam House, but he advised me to join one of the colleges instead. David Shoenberg, who was already a research student and who later became a well-known solid-state physicist, suggested to try Trinity, St. John's, and Magdalene; but there was no space available at Trinity and St. John's would only let me know much later. When I tried Magdalene College, the Senior Tutor, V. S. Vernon-Jones, looked at me, realizing that I was a refugee and said 'I suppose we ought to have one.' He also offered me financial support. I stayed at Magdalene through my PhD studies, and then for an additional two years as a Junior Fellow (Charles Kingsley Bye-Fellow).

In principle, Rutherford was in charge of all the students, but he would distribute them according to their interest. In 1933, I still wanted to be a theorist. He suggested that I work with Ralph Fowler, his son-in-law, who was the theoretician at the Cavendish Laboratory. Fowler let you alone if you knew what you wanted to do. His former students included Dirac and Chandrasekhar. With Dirac, he gave him a preprint from Heisenberg and that was enough."[2]

In view of Goldhaber's expertise in nuclear physics, it seemed puzzling that he was not included in the Manhattan Project—by the time this project commenced, he had become a renowned authority. He told us about what happened.

"There were various excuses, but it came out indirectly. One man early on who was in charge was a theoretical physicist, [Gregory] Breit. I saw him in 1940 during a

---

[2] Ibid., 216–17.

meeting in Washington. He asked me what ideas I had and I told him some ideas. He literally threw up his arms and said 'But Fermi and Szilard had the same ideas.' My feeling was that they thought that they had already enough of these 'important' foreigners. We worked at the University of Illinois at the time; the Manhattan District followed my neutron work carefully, and they asked for my permission to copy my neutron counters. They were then used at the first pile in Chicago. Breit acted as a censor. When I wanted to publish a paper, they would send it to him from *Physical Review*. Then the editor would write me a polite letter whether we would be willing to leave the paper unpublished during the war, and, of course, we did. We had about half a dozen papers that were published after the war. However, our data were used in the project although there was never an acknowledgement for it. In 1943, Breit wrote to me saying that it's a pity they couldn't use us because my wife's parents were still in Germany. But there were others in a similar situation who worked on the bomb. This was just an excuse. I regret very much that I was not aggressive enough to barter my expertise for getting Trude's parents a visa to the United States. The Nazis killed my wife's parents already in 1941 and we know the exact day when they did it, because the Germans kept good records!"[3]

---

[3] Ibid., 221.

DAVID GROSS

You have to know a lot to ask intelligent questions.

David Gross and Magdolna Hargittai in 2005 in Lindau, Germany. (photo by I. Hargittai)

David Gross[1] (1941–) was born in Washington, DC. He received his bachelor of science degree in 1962 from the Hebrew University of Jerusalem and his PhD degree in 1966 from the University of California at Berkeley. He spent twenty-seven years at Princeton University and is now the Frederick W. Gluck Professor of Theoretical Physics at the University of California, Santa Barbara. He was awarded the 2004 Nobel Prize in Physics jointly with H. David Politzer

[1] I. Hargittai and M. Hargittai, *Candid Science VI*, 838–55.

and Frank Wilczek "for the asymptotic freedom in the theory of the strong interaction."[2] We recorded our conversation in 2005 in Lindau, Germany.

On the comparison of research in fundamental physics and molecular biology.

"I don't think that it's a question of competition, say between fundamental physics and biology, because there is room for everybody. It's not that you can save money by not doing high-energy physics and then spend it on biology. The amount of money being spent in biology is at least by a factor of ten more than all the physical sciences put together. So it's not a question of money. I don't know what will happen if we get to the point where there are fascinating questions, which interest everyone, that we cannot afford to study. Also, it's hard to predict what technology will make use of fifty years from now. In the past we have found applications for developments in basic physics, which, at the beginning, seemed totally removed from everyday life, like electricity. More recently, we have the example of quantum mechanics—pure science in the 1920s, today the foundation of modern technology. So who knows what will happen. I have a feeling that it would be a momentous and tragic moment in human history if we get to the point when we aren't able to answer deep questions like the origin of the universe because they are too big and too expensive to attack. Actually, I'm—for no good reason—very optimistic that we'll find solutions for these problems that today may seem hopeless.[3]

Could the human mind go to deeper understanding than at the present level of frontier science?

This is a question that I often discuss in talks; it's about whether there is a limit to our capability to understand the universe and its laws. I have a feeling that there is no such limit. I like dogs, but I know very well that I can't teach my dog quantum mechanics. Not only can't I teach it quantum mechanics, I can't even teach it classical mechanics. There's a limit to what a dog can understand. Is there a limit to what the human mind can comprehend? Does the universe contain levels that are beyond human understanding?

I believe strongly that the answer is no; we are capable of understanding everything. I believe this for a variety of reasons. One is that language has an infinite capacity. Noam Chomsky made the famous remark that one of the most important characteristics of human language is that even a baby can formulate an infinite number of sentences once it acquires a small vocabulary. In other words, a baby has an innate ability to be able to formulate new sentences that have never been formulated before. There's something about language, and the higher form of language called mathematics, that suggests an infinite capability. I see no evidence in the structure of knowledge—that has developed over the centuries—of reaching boundaries, of facing problems that we would not be capable of understanding.

---

[2] "The Nobel Prize in Physics 2004," *Nobelprize.org*. Nobel Media AB 2013, http://www.nobelprize.org/nobel_prizes/physics/laureates/2004/.
[3] *Candid Science VI*, 853–54.

Finally, I have empirical evidence. If it was true that the world, beyond some point, is impossible for us to understand, if there was a limit to what we could understand, then surely it would've been the case that it would take longer and longer to educate young people to get to the point where they could start making contributions to science. But that's not the case. It takes about the same amount of time for the brightest minds to get to the point when they can shake the world as it did five hundred years ago. Newton made his contributions in his twenties. Today we're dealing with much more complex things, but it takes young people no longer to get to the frontiers of physics. I don't see any sign that we're getting anywhere near a limit to our ability, and until there is evidence to the contrary, I will continue to believe that there is no such a limit."[4]

---

[4] Ibid., 854–55.

ANTONY HEWISH

I couldn't have been a more lucky astrophysicist.

Antony Hewish in 2000 in Cambridge, UK. (photo by I. Hargittai)

Antony Hewish[1] (1924–) was born in Fowey, Cornwell, England. He is Professor Emeritus of Radioastronomy at the University of Cambridge. He was elected a Fellow of the Royal Society (London) in 1968. Professor Hewish received the 1974 Nobel Prize in Physics jointly with Martin Ryle for their results in radio astrophysics. Hewish's prize was specifically for his decisive role in the discovery of pulsars. We recorded our conversation in 2000 at the Cavendish Laboratory.

This is what Hewish told us about the prize-winning discovery.

"I designed a radio telescope to make a survey of the sky to detect radio galaxies of smaller angular size because they would be the ones that showed this diffraction effect most. It needed to be very sensitive because I wanted to see a large number of galaxies. It also needed to measure this scintillation effect, which shows up best—because it's related to the plasma in the solar wind—at longer wavelengths, which radio telescopes

---

[1] M. Hargittai and I. Hargittai, *Candid Science IV*, 626–37.

were not then using. I had to design a special purpose instrument to study this. It had to work at meter wavelength. It had to be a huge area because I needed high sensitivity, and for my method it was necessary to make repeated measurements on radio galaxies to get their angular size.

The survey was set up to observe several hundred radio galaxies every week. Now, by a very lucky accident, the instrument I designed was ideal for detecting a completely unknown phenomenon called the pulsar. It was totally unexpected, unpredicted, and one of those shattering things that science brings up, like the discovery of X-rays. We couldn't have avoided detecting the pulsars. These are sources, which produce regular flashes of radiation about once a second or sometimes much faster. This phenomenon was totally unanticipated. When we started to see these flashes in November 1967, it took about a month of extremely hard work before I decided it was probably neutron stars, but I wasn't quite sure when we published the discovery in *Nature*. It was that discovery which led me to the Nobel Prize.

The prize was awarded for what they said was my decisive role in this discovery. What that means was that I conceived the project, designed the apparatus, and decided what observations were to be made. Other people were actually involved in carrying it out for their PhDs and so on. It turned out to be the most amazing, fruitful discovery.

Did you accomplish you original goal too?

Oh, yes, but that took much longer. I did that as well. We mapped all these galaxies, and that involved several research students over the next decade. It was quite a long program but the pulsar just came out immediately. It was beautiful. I couldn't have been a more lucky astrophysicist.

Do you think you would have received the Nobel Prize without the pulsars?

No.... You've got to do something very special to get the Nobel Prize.

Was Martin Ryle's discovery primarily methodology?

No, not just that. They mentioned 'pioneering work in radio astrophysics' in both our cases. Martin Ryle's invention of the aperture synthesis revolutionized the design of radio telescopes. Everywhere today you find his technique employed. But his observations gave the first evidence for the 'Big Bang' universe, as opposed to the steady state theory, which was fashionable then.

Didn't you have a particular graduate student who made the first observation of pulsars?

Oh, yes, I did. She was my student doing observations, which I had designed. She was a good student and worked very hard, helping to build the radio telescope and then carefully analyzing hundreds of feet of chart recordings.

What is her name and what happened to her?

She is Jocelyn Bell Burnell; Bell is her family name and Burnell is her married name. She has stayed in the field and after a variety of posts, as she had to follow her husband's jobs at the beginning, she has now a chair in physics (astrophysics), at the Open University in Milton Keynes. She has done very well for herself. There are not many women in this country who have a chair in physics."[2]

---

[2] Ibid., 631–33.

GERARD 'T HOOFT

Nature outsmarts people.

Gerard 't Hooft in 2001 in Utrecht, Holland. (photo by M. Hargittai)

Gerard 't Hooft[1] (1946–) was born in Den Helder, Holland. He is Distinguished Professor of Physics in the Institute for Theoretical Physics and member of the Spinoza Institute of Utrecht University. He obtained his PhD degree from the University of Utrecht in 1972 and has been there ever since. He received the 1999 Nobel Prize in Physics together with Martinus Veltman "for elucidating the quantum structure of electroweak interactions in physics." [2] He has received numerous other honors as well. We recorded our conversation at the University of Utrecht in 2001.

---

[1] M. Hargittai and I. Hargittai, *Candid Science IV*, 110–41.
[2] "The Nobel Prize in Physics 1999," *Nobelprize.org*. Nobel Media AB 2013, http://www.nobelprize.org/nobel_prizes/physics/laureates/1999/.

We asked Gerard 't Hooft whether he thought there was such a thing as a Grand Unifying Principle or a Theory of Everything?

"As soon as the Standard Model was more or less clear on paper—its equations, its peculiarities— it was clear to everybody who studied it that there was a grander pattern in it; it's not just random. There appear to be quarks and leptons, and they show much more of a pattern than what we'd expect if the creator of the universe just took a bag of rubbish and put it all together, saying 'this is the model.' It is not that. There is an enormous amount of systematics in the Standard Model. It was also clear to everybody that there must be further explanations for why the model is built this way: there must be an explanation of the structure; there must be a further unifying principle. It's much like what the chemists must have discovered when Mendeleev came along with the Periodic Table of the Elements. It was quite clear that there was a pattern in this; it was not just arbitrary. But the chemists of that time could not understand what made the system work; they only recognized the pattern. Here we are now; we see the Standard Model and we see that there is a system in it. It's obvious to anybody who looks at it that there must be a more unified theory that explains this particular pattern that we see in the Standard Model. Yes, we all expect that there will be a more unified description, and I have sufficient trust in human ingenuity that people will figure out...what the deeper reasons are for this particular pattern that we see in the Standard Model. But we haven't figured it out yet—although there are many ideas and suspicions.

Will it be a simple explanation or a complicated one?

I think it will be a surprising explanation. It happens all the time that nature outsmarts people. I expect that it will be a beautiful smart argument or reason; and I'm also sure that once people put all this together many of us will say, we should have figured this out long ago; it is so simple! But we just did not think of it. Now it looks very complicated and mystifying to us.[3]...

Do you believe in having human colonies in the future, outside Earth?

Yes, I like to dream about that. In principle there is nothing standing in our way. Particularly, the Moon itself is within reach. From purely technical point of view, there is nothing against colonizing the Moon. The only real reason not do it yet is that, even on Earth itself, there are so many places that could be colonized but have not been colonized yet. The deserts are still empty; in Canada, there are enormous territories empty. There is no reason why those places are uninhabited other than purely economic ones. It is not yet worthwhile to live there because life is harsh; and you cannot get a good job, it is just uneconomical. To inhabit the Moon will be much more difficult than to live in any desert or tundra on the Earth.

Do you believe in extraterrestrial intelligence?

No. There are two questions to be asked. One is whether there exists an extraterrestrial intelligence, and the other is whether it is possible to make any contact with them or whether there has already been some contact with them. That second question, I think, is to be answered negatively, they are probably just too far away. There might

---

[3] *Candid Science IV,* 121–22.

be a one-way contact in a sense that one day we'll find some signals—it's not totally inconceivable; maybe we can even send signals—but we won't be able to communicate because they will be too far away. Even as to the question of their existence; I think that very often people are just far too optimistic about the possibility of intelligent life arising anywhere. I find it extremely improbable, though not impossible, that there is life anywhere inside the solar system other than on the Earth. Some people think that there might be life on Jupiter's moons, or on Mars; I find that very improbable. We have seen it on Earth that although the conditions here were extremely hospitable and in spite of all the good chances, life developed extremely slowly. Most of the other planets in the solar system are much less favorable for life. Yes, there are probably planets orbiting distant stars that are favorable for life but their number would be very difficult to estimate. People appear not to understand the implications of large numbers. This question is determined by many large numbers that might cancel each other out. There is a large number of planets in the universe, and you can estimate this number. There are large numbers describing how probable or improbable it is for life to emerge. Those large numbers are expected to cancel each other out, but there is no reason for them to cancel out completely. It is possible that there are billons of planets with intelligent life like us or maybe we are a one out of a billion chance for the entire universe, it could be either way. I am a bit pessimistic and I would guess that this Earth is a very very lonely place. Maybe there are entire galaxies without any sign of life whatsoever."[4]

---

[4] Ibid., 137–38.

WOLFGANG KETTERLE

I gave him the key to the family car because I knew he could drive it faster than I could've driven it. Ketterle quoting David Pritchard

Wolfgang Ketterle in 2002 at MIT. (photo by M. Hargittai)

Wolfgang Ketterle[1] (1957–) is John D. MacArthur Professor of Physics at the Massachusetts Institute of Technology. He was born in Heidelberg, Germany, and received his education in Munich, including his doctorate in 1986. Following postdoctoral studies in Heidelberg and at the Massachusetts Institute of Technology (MIT), he joined the faculty of MIT in 1993. Ketterle together with Eric A. Cornell of the Joint Institute for Laboratory of Astrophysics (JILA) in Boulder, Colorado, and Carl E. Wieman, also of JILA, received the 2001 Nobel Prize in Physics "for the achievement of Bose-Einstein Condensation in dilute gases of alkali atoms, and for early fundamental studies of the properties of the condensates."[2] We recorded our conversation in 2002 at MIT.

---

[1] M. Hargittai and I. Hargittai, *Candid Science IV*, 368–89.
[2] "The Nobel Prize in Physics 2001," *Nobelprize.org*. Nobel Media AB 2013, http://www.nobelprize.org/nobel_prizes/physics/laureates/2001/.

Dan Kleppner, Wolfgang Ketterle, Thomas Greytak, and David Pritchard at MIT. (courtesy of W. Ketterle)

A summary of Bose-Einstein Condensation

"Bose-Einstein condensates can be regarded as the coldest matter that man has ever produced. It may even be the coldest matter in the universe. Matter at those temperatures has very special properties. It behaves like no ordinary matter. The special properties of this very special matter can be best described by using a comparison with the optical laser. If you compare a light bulb to a laser beam, the difference is very striking. The light bulb emits light in all directions, that is, many electromagnetic waves go from it in all directions. In the laser beam, the light is directional; it's just one single wave. There is a similar difference between an ordinary gas and the Bose condensate. In an ordinary gas, the atoms move around independently, in all directions. In the Bose condensate, all the atoms march in lockstep, it just forms one big wave."[3]

How did it happen that Ketterle 'inherited' a project from David Pritchard?

"I liked MIT and I appreciated Dave Pritchard's mentorship, and Dan Klappner's too, so I decided to stay. I still can't fully understand what Dave did for me; it is unprecedented. Dave handed me over his lab and the laser cooling apparatus, and told me just to go and run with it. He told me that I could still ask him for any advice, but I was independent to make all decisions, and that we wouldn't publish the results together. He gave me full independence, but he also gave me a head start. It was unbelievably generous. When people start as an assistant professor and build up their own lab, if they are fortunate, after two years they may have built up some complicated machinery

---

[3] *Candid Science IV*, 370–71.

and may publish their first paper. I was an assistant professor for two years and I was the codiscoverer of the Bose-Einstein condensation. I know that I couldn't have done it without the support of Dave Pritchard. In fact, when we did accomplish the Bose-Einstein condensation, I invited Dave to be a coauthor of our first publication, because I regarded it the culmination of our previous collaboration, but he decided to stay off it.[4]

It seems as if he had just stepped out of the most exciting part of it.

[Pritchard] once used this expression, 'I gave him the key to the family car because I knew he could drive it faster than I could've driven it.' I have never heard any word of regret from him. He has always taken a fatherly pride in my accomplishment. He knows that he has done a lot for MIT and he has done a lot for atomic physics, and he has done a lot for me. The leader of the first group that realized Bose-Einstein condensation, Eric Cornell, was his student; and the leader of the other group was his postdoc. So he influenced the field in a major way.

He must be an exceptional person, because the Nobel Prize is the ultimate recognition in science and having no regret seems to be superhuman. Was your goal winning the Nobel Prize?

[After a long silence] You see me hesitating…because the Nobel Prize was never my goal. My goal has been excellent science, which I have enjoyed doing. Just to be in the midst of exciting results, living through such an exciting time was ample reward. The Nobel Prize recognizes that, and it needs a combination of luck and other things. I would never recommend to any scientist to aim for the Nobel Prize and I did not have it in my mind, either. There is more Nobel Prize-worthy work than can be rewarded and the quality of my work would be the same if I hadn't gotten the Nobel Prize for it."[5]

Ketterle's relationship with Dr. Pritchard was part of a long chain. Dan Kleppner was Norman Ramsey's pupil and Kleppner was at one time Dave Pritchard's mentor.

"I am very close to Dan. The pursuit of Bose-Einstein condensation of diluted atomic gases started with working on polarized atomic hydrogen in the late 1970s. The leading groups were a group in Amsterdam and a group at MIT led by Tom Greytak and Dan Kleppner. It was their efforts that put Bose-Einstein condensation into the minds of researchers. For a whole decade this work focused on atomic hydrogen gas. In the end of the 1980s, with the advent of laser cooling, there were different atoms that could now be used to pursue ultracold atomic physics. But the original work on atomic hydrogen directly effected or even triggered the search for Bose-Einstein condensation in the alkali atoms. I knew all the preceding work, I had read all the papers, and it was an enormous source of inspiration. For me, there was even more direct inspiration because Dan Kleppner is here at MIT, his office is in this hallway, and we often talked to each other."[6]

---

[4] Ibid., 375.
[5] Ibid., 376–77.
[6] Ibid., 377.

NICHOLAS KURTI

I thought to myself, why should I ever leave this place [Oxford], and I never did.

Nicholas Kurti in 1994 in London. (photo by I. Hargittai)

Nicholas Kurti[1] (1908–1998) was born in Budapest as Miklós Kürti. After completing high school in 1926, at the age of eighteen, he left Hungary because of the existence of anti-Jewish legislation. He studied in Paris and Berlin, worked in Breslau, Germany (now the city of Wroclaw in Poland), before the Nazis forced him out of Germany; and he settled in England. He was Professor of Physics of the Clarendon Laboratory of Oxford University, Fellow of the Royal Society, and Commander of the British Empire. We recorded our conversation in 1994 at the headquarters of the Royal Society in London.

The road from Budapest to Oxford.

"At the age of eighteen, I went to Langevin in Paris. Fortunately, I had learned French by then. Langevin told me that first of all I must get a *Licence ès sciences physiques*

---

[1] I. Hargittai and M. Hargittai, *Candid Science VI*, 554–65.

degree [a college degree majoring in physics]. This meant passing a rigorous examination, first a written test and then an oral one. I succeeded doing this within two years. I studied chemistry, physics, and mathematics. I had the most famous professors in both chemistry and physics. For the continuation of my studies Jakab Szentpéter's [a physicist friend in Budapest] idea was to go to Berlin. For this, I had a letter of introduction to Michael Polányi, who was at that time in Berlin. Polányi suggested to me to do one year of postgraduate work and then to do a doctorate. The field I chose was low-temperature physics, and Professor Franz Simon was my supervisor. He was one of the founders of low-temperature physics in Germany. Those three years, between 1928 and 1931, in Berlin were the most fantastic. As a city to live in, Berlin did not appeal to me....

The most important thing, though, was the *Physik Kolloquia*, organized by Max von Laue in the physics department. These were not colloquia in the present sense of the word. They were more like the American journal clubs, just one two-hour session every Wednesday. A few people simply reported on recent publications from the literature. It was characteristic that in 1929 or 1930, Max von Laue could have an overview of the whole physics literature by looking at the *Proceedings of the Royal Society, Physical Review*, and *Physikalische Zeitschrift*. If you went regularly to this colloquium, you could know what was going on in physics. Then you could keep up with everything. Laue would ask the audience about papers as he was looking for volunteers to review them for next time. It was regarded as the thing to do for graduate students to volunteer. Just think of it. You were reporting about a recent paper by a famous physicist and there was the audience, in the front row, Planck, Schrödinger, von Laue, Gustav Hertz, Haber, Nernst, about six or seven Nobel laureates or future Nobel laureates. Behind them were Wigner, Szilard, and others. It was a very interesting experience. It was also wonderful to see that every now and then, the great men could also make some silly mistakes. I remember once when Schrödinger suddenly stood up in the middle of a discussion of the spectra of triatomic molecules and suggested that the calculations could be simplified if you assumed that the three atoms were in the same plane. There was a silence, followed by laughter.

I was doing low-temperature physics and almost blew up the laboratory. I was the first person to produce very strong magnetic fields by magnet coils cooled in liquid hydrogen. I was doing my first experiment, and everything looked all right and I pressed a button, which sent a current of 40 amps through a copper wire of half a millimeter diameter. So I pressed the button and there was a huge bang and my Dewar blew up. Fortunately, I was standing about two meters from it and nothing happened to me. What had happened was that I had this wire in liquid hydrogen, and in order to save hydrogen, I precooled it with liquid air. At that time, we didn't have liquid nitrogen, just liquid air. My mistake was that I didn't wait until all the liquid oxygen evaporated. There must have been a spark, a bad contact, and that did it. Eventually, I got my experiment right and got my doctorate too.

Then I had a private assistantship with Simon. In the meantime, he got a full professorship in Breslau [then, in Germany, now Wroclaw in Poland] at the Technical University, so I followed him there. Then Hitler came to power and Simon decided that it was not the place to remain, especially since he had two small children. Although Simon was Jewish, at the beginning, he was not subject to the anti-Jewish law as he had served in World War I and was awarded the Iron Cross, First Class. He used to mention many times that he was among the first gas casualties on the Western Front. Simon knew F. A. Lindemann [later Lord Cherwell], and it was arranged that Lindemann would invite him to England, and he brought me with him as his assistant. This was in the spring of 1933. Simon decided not to make a big fuss out of his departure. He just resigned his Chair. He called me into his office and told me that he just signed his letter of resignation when he received a letter from the *Notgemeinschaft der Deutschen Wissenschaft* [Emergency Association of German Science], which was the German funding agency for scientific research that time. Simon had applied for a big grant six months before. Now he received this letter telling him that all his requests had been accepted, including the complete refurbishing of his laboratory. Simon left nevertheless. This is how I came to Oxford."[2]

---

[2] Ibid., 557–58.

BENOIT B. MANDELBROT

My long scientific life defied and survived categorization.

Benoit B. Mandelbrot in 2000 in Stockholm. (photo by I. Hargittai)

Benoit B. Mandelbrot[3] (1924–2010) was born in Warsaw, Poland, and became most famous for his pioneering work on fractals. In 1936, he and his family moved to France, where he survived World War II in spite of anti-Jewish persecution. He graduated as an engineer from the École Polytechnique in Paris in 1947; a year later he earned his master's degree at the California Institute of Technology, then worked for the Dutch Philips Company, and in 1952 earned his doctorate at the Faculté des Sciences de Paris. From 1958 he worked at IBM in the United States. Referring to his revolutionary ideas of introducing the fractal description of everything, the citation of one of his highest recognitions, the Wolf Prize for Physics (Israel, 1993), stated that he has changed our view of nature. Mandelbrot had a strong mother who had expected great things of her son. In Mandelbrot's words, "She expected much of me, and I was slow in developing. She lived long enough to be reassured, I think, but without this

---

[3] M. Hargittai and I. Hargittai, *Candid Science IV*, 496–23.

extraordinary piece of luck of bumping into fractals, I might have amounted to nothing compared with her expectations."[4] We recorded our conversation in 2000 in Stockholm.

On the birth of the concept of fractals.

"There was a gradual process that began by exclusion. I had not liked Caltech or Philips engineering and decided to return to science and obtain a PhD. One possibility was theoretical physics. However, there was no role model in France; and besides, I did not want to work on anything related to the atomic bomb. I wanted to do geometry of the classical kind, but my uncle had convinced me that this topic was at a dead end. I hated algebra, so straightforward mathematics was out. I was an ambitious and rebellious idealist looking all around for some domain in a terrible mess to which I might try to bring mathematical order. All the while, my uncle was joking that my ambition was to become Kepler, and I was late by several centuries.

But I did not think so because of something I had noticed. As the disciplines become organized, there is an increasing accumulation of observations that fit nowhere. I was free to look for potential treasure in dumps hidden in dark corners. Eventually, and against all odds, I managed to clear several terrible messes, and my romantic dream was completely fulfilled. For a person to be entitled to say so is truly overwhelming. An old man saying so is necessarily filled with awe. The process took a very long time, and it is only recently that I began to see a system in my wild search. I realized that the fields I had excluded can be said to be ruled by smoothness, and my life work was to provide a theory of roughness.

When you gave the world this new word 'fractal,' did you think long about it?

A text I wrote in 1975 became the first book on fractals, but its original title was very clumsy. I gave the draft to Marcel-Paul Schützenburger, a dear friend and an extraordinarily brilliant man, who knew French thoroughly. He was a genius, but too much of a dilettante and so disorganized that he is already forgotten. The first thing he told me is that if I would try to peddle this ,,,, (he used a French five-letter word), he would protest to my publisher because he knew I could write in proper French and therefore must do so. I reread my work and realized that 'Marco' (as we all called him) was right. That draft was full of Anglicisms. Someone had to help restore my previous command of the language.

The second thing Marco said was that I had brought to life something that did not exist before, so I was entitled to give it any name. Roughness is a concept connected with fracture so I looked up a Latin dictionary and was reminded that the word 'fracture' comes from the Latin adjective *fractus, fracta, fractum*, which means 'like a broken stone.' The Romans did not care much for abstraction; they used very concrete words. The proper neologism was the Latin word 'fractum,' but it rhymed with quantum. This made it pretentious, in bad taste. Next, I hit on 'fractal,' which sounded good

---

[4] Ibid., 500.

in French and English—and later in many other languages into which my book has been translated. It is now in every dictionary.

Of course, there were various linguistic difficulties. In Russian, should the last letter of 'fractal' be hard or soft and what should be the gender? A committee decided that it should be the soft fractal' and that it should be feminine. If you compare new scientific terms, 'fractal' is less problematic than 'chaos'. Every old word continues to be burdened by its previous meanings, but a new word has none. Of course, my word, fractal, has already attracted new meanings.

So what is the mathematical definition of fractals?

There is no single mathematical definition. However, even in mathematics everything that is important is difficult to define. If you don't believe me, try to define probability theory. Or note that 'general topology' is defined as 'the study of the notion of neighborhood'. For general purposes, I follow this example and call fractal geometry 'the study of roughness'.

If pressed further, I add that it is a theory of self-similarity and related invariances. A self-similar shape is one that looks the same from nearby and afar. Every formal definition I know excludes something very important, and I no longer look for a single all-encompassing definition. My first book gave no definition. Then I relented and gave a tentative definition based upon fractal dimension. But a prominent French mathematician, a formalist and a snob, criticized me sharply. He understood fractality and viewed it as a very fundamental notion. To the contrary, fractal dimension is a specialized notion that has many variants. Therefore, he told me, I should not define a fundamental notion on the basis of a specialized one. Physicists think differently. Fractal is a concept but fractal dimension is a number that quantifies roughness and can be measured empirically. Physicists detest concepts and worship numbers that can be measured.

An important question is whether or not fractal geometry will survive, whether or not it will continue to provide good service to science and economics. But this question has nothing to do with whether we can find an all-embracing definition."[5]

---

[5] Ibid., 507–9.

RUDOLF MÖSSBAUER

We learn that the Sun works the way we thought it does.

Rudolf Mössbauer in 1995 in Budapest. (photo by I. Hargittai)

Rudolf Mössbauer[1] (1929–2011) was a German physicist. He was born in Munich, graduated from the Technical University of Munich, and in 1958 received his doctorate from the same institution. He worked at both the Technical University of Munich and the Max Planck Institute in Heidelberg. At the Planck Institute he discovered the existence of recoilless nuclear resonance absorption, known as the Mössbauer effect, on which Mössbauer spectroscopy is based. In 1960 he moved to the California Institute of Technology. The following year he was awarded the 1961 Nobel Prize in Physics for his discovery. In 1965, he returned to Germany to become a professor at the Technical University of Munich, where he stayed until his retirement in 1997. Between 1972 and 1977, he was director of the Institut Laue-Langevin in Grenoble, France. We recorded our conversation in 1995 in Budapest.

---

[1] M. Hargittai and I. Hargittai, *Candid Science IV*, 260–71.

Mössbauer's major research interest in his post-Mössbauer-effect period was about neutrinos. He explained to us why he found this research important.

"For the first time we learn that the Sun works the way we thought it does. We used to suppose, and now we know it, that the energy of the Sun comes from nuclear fusion. Of course, the same is true for all the other stars as well. It is basic research, and there are no immediate applications for everyday life, but there is an enormous amount of side products."[2]

Mössbauer was actively involved in the GALLEX (Gallium Experiment) Project.

"We measure solar neutrinos in this experiment, which is carried out by a huge international collaboration. The neutrino is a particle which carries no charge and which is subject to weak interactions only. Only the gravitation is still very much weaker; but it matters only on astronomical scales. Our body, for instance, is attracted by the entire Earth, and it is only for this reason that gravitation appears significant to us. On nuclear scales, gravitation is negligibly small, even when compared to the so-called weak interaction. This latter interaction is so weak, that a neutrino entering the Sun on one side and leaving it on the other side hardly feels its existence. Neutrinos, in particular, are formed inside the Sun, as they are in other stars, when four protons are being fused into a helium nucleus. The helium nucleus is a little lighter than the four protons, and this mass difference is converted into the energy which the Sun radiates. The generated neutrinos then leave the Sun, some coming to the Earth where we then can measure a tiny fraction of them.

If the neutrinos can travel through the Sun, how do we know whether they have traveled through the Sun or were generated in the center of the Sun?

They can only stem from the Sun. The other stars produce similar amounts or even more neutrinos, but they are so far away that the solid angle which the Earth forms to them is so small that we cannot observe them. Only the Sun is sufficiently close so that we can measure its neutrinos.

How do you detect them?

It is simple in principle and complicated in practice. You use a big volume of a proper material. Take an ordinary radioactive decay, which goes from a higher energy state to a lower energy state, and you take the inverse. You shine the solar neutrinos on a nucleus and occasionally you get from the ground state of this nucleus with atomic charge Z to a higher state in a nucleus with atomic charge Z + 1. You need a minimum neutrino energy for this transition, giving rise to an energy threshold. If there is more energy available, then that's all right. Gallium is especially favorable for this experiment because, among other things, relatively little energy is needed to initiate the transition to the neighboring element, which is germanium. The energy threshold, therefore, is low, and thus one can detect the major part of the solar neutrino flux. Only a few of the incident solar neutrinos perform the transition from the ground state in gallium into the higher state in germanium. The measurement is then being done by looking at

[2] Ibid., 267.

the inverse transition back, germanium to gallium. The number of such back decays is representative of the number of initial excitations and thus of the number of incident solar neutrinos. This is how their number is being determined. Because we are dealing with weak interactions, most neutrinos go unnoticed through the detector, and there is only occasionally an interaction, about one per day."[3]

---

[3] Ibid., 265–66.

YUVAL NE'EMAN

A society which tries to regulate its research people and requires them to keep to planned utilitarian targets is condemned to stagnation.

Yuval Ne'eman in 2000 in Stockholm. (photo by I. Hargittai)

Yuval Ne'eman[1] (1925–2006) was a native of Israel. He had parallel careers in the military, politics, and science. He made Nobel Prize-level discoveries in the field of fundamental particles physics; yet, according to his own estimate, he devoted only about half of his time to research. Murray Gell-Mann and Ne'eman both contributed decisively to the creation of a system of fundamental particles. Gell-Mann, though, devoted his entire career to science, and he was also very good at giving appealing names to his discoveries; Ne'eman gave dry names to his. In 1969, Gell-Mann was awarded an unshared Nobel Prize in Physics "for his contributions and discoveries concerning the classification of

---

[1] M. Hargittai and I. Hargittai, *Candid Science IV*, 32–63.

Murray Gell-Mann in 2001 in Stockholm. (photo by I. Hargittai)

elementary particles and their interactions."[2] The citation could have applied to Ne'eman as well. Ne'eman was a wunderkind in high school. He received his bachelor of science degree and his master's degree in physics from the Technion–Israel Institute of Technology in 1945 and 1946, respectively. It was another sixteen years before in 1962 he earned his PhD degree at London University. We recorded our conversation in 2000 in Stockholm.

Ne'eman made predictions concerning the Higgs particle.

"My model predicts that the mass of the Higgs is twice the mass of the W, which is 85 GeV, so it will be 170 GeV. However, there is a renormalization correction because at high energy the mass is a function of energy. The energy at which this Pythagorean result holds is a higher energy. We calculated the mass of the Higgs at lower energy and it comes out as 130 ± 10 GeV. My prediction is the only prediction in the field. No other theory says anything about the Higgs—except for ordinary supersymmetry, which then requires the existence of lots of new particles....

If and when the Higgs is found and its mass measured, I would like to advertise my theory for people to know that I had predicted it. I have suggested to my co-inventor, David Fairlie, that we should collect our reprints in a volume, for example, or write a review in *Physics Reports*. David is still worried about some things and wants to calculate some quantities to make it easier to understand our approach. The name of the supergroup is SU(2/1). It is now all published material; it is all there, and I only want people to know about it."[3]

---

[2] "The Nobel Prize in Physics 1969," *Nobelprize.org*. Nobel Media AB 2013, http://www.nobelprize.org/nobel_prizes/physics/laureates/1969.

[3] *Candid Science IV*, 50–51.

He spoke about science in Israel, which was already substantial, but it was still before the several Nobel Prizes that were in a few years' time awarded to Israeli scientists.

"In 1983, when I was the first Minister of Science I was invited to the exclusive Elsevier "Economist" Club to give a lecture about Israel's science-based industries. This club meets once a month in Haarlem, Holland. Before me, the lecturer was Helmut Schmidt, the former Chancellor of West Germany and the one after me was Henry Kissinger. We had strong science-based industries in 1983, and since then they have further developed well. They are spread all over the country. They started many years ago next to the Weizmann Institute and next to the Technion. Today there is also a large concentration in the Tel Aviv area and in other places. They employ people who left the universities, among them scientists who became successful entrepreneurs. As a matter of fact, our key economic strength is in 'high tech.'. .

In present-day Israel, the computer science groups in industry are very strong. There we have a large proportion as compared to America and Europe…although there are weaknesses. Generally, we produce 1 percent of world science, although we are 0.1 percent of the world population. The scientific infrastructure is weak. Theoretical physics is very good, there are excellent people in string theory, for instance. Experimental physics is much weaker because not enough is spent on infrastructure. There are highs and lows. I tried to correct it at the time but now it is the opposite.

The Russian immigration helped our physics. There was a decrease in physics graduate students, which was reversed by the immigration. This was so for the other sciences as well. Now they tend to go to business, like everybody else."[4]

About the Jewish contribution to science.

"The graduate schools in America in physics used to be full of Jews, and now they are full of Asians. I have my explanation about Jews in science. The Jews were out of society until about 1800 and were allowed only to be peddlers. On the other hand, at least all males could read and write. Then came two revolutions, one was the French Revolution, which started Jewish emancipation, and the other was the Industrial Revolution, which created new types of white-collar jobs, such as engineers, accountants, lawyers, and other new professions. European society was generally traditional, with considerable inertia; the sons often following their fathers in their professions. So the new professions were not very popular. On the other hand, the Jews were allowed, just at that time, to move from peddling to other professions. They were also better prepared than many others, with their background in education. They jumped at the new opportunities, and by 1900, especially in Central Europe, a large proportion of white-collar professionals were Jewish. That still did not mean science, which was still mostly the privilege of the previous order. Even though the impact of religion was gradually waning, many of the Jewish professors, when they were moving up the academic ladder, had to convert. There are perhaps also other factors. The study of the Talmud prepared the Jewish scientists for abstract thinking. I even have

---

[4] Ibid., 54–55, 56.

a conjecture about a genetic component. Throughout two thousand years, there was pressure to convert. Those who resisted the pressure were those who were more apt than others to create and live in an abstract world of their own. That may have created some selection. There were more Jews in theoretical areas and less in experimental and engineering fields.

A Dutch author has a book on Jews in science, with a different genetic explanation. According to him, the Christian world for many centuries, between Constantine and the Reformation and in Catholic communities even longer, bred itself out of talent for abstract thinking. If there was a son who was more theoretically inclined, he was sent to the church and this was in practice a way of getting the genes out. The Jews had the opposite effect. A rich Jew wanted to marry his daughter to a scholar, and they would live at his table, and this would breed more scholars."[5]

---

[5] Ibid., 55–56.

MARK OLIPHANT

The theoretical physicist is the closest to God.

Mark Oliphant in 1999 in Canberra, Australia. (photo by I. Hargittai)

Marcus (Mark) Laurence Elwin Oliphant[1] (1901–2000) was born in Kent Town, near Adelaide, South Australia. He was co-discoverer of tritium and helium-3 with Ernest Rutherford; principal coworker in establishing efficient radar for the defense of the Allies in World War II; a leading British participant of the Manhattan Project; founder of the Australian Academy of Science; and co-founder of the Australian National University. We visited him on July 22, 1999, at his home in Canberra, where he lived with his daughter, Vivian Wilson. At the age of ninety-eight, he was still a commanding presence. We came away from this visit deeply impressed by his wit and care and interest.

---

[1] M. Hargittai and I. Hargittai, *Candid Science IV*, 304–15.

Lord Rutherford demonstrates the deuterium reactions at the Royal Institution in 1934. Oliphant is standing in front of the bench. (courtesy of the late Mark Oliphant)

Recalling his most important discovery.

"We were doing experiments with all the possible projectiles in order to produce transformations in elements. It was natural to try to use heavy hydrogen and, indeed, the results were very interesting. The experiments with heavy water brought about two discoveries; one was helium-3 and the other was tritium."

On his interactions with Rutherford in connection with these discoveries.

"Rutherford was the greatest influence on me and on so many other people at that time in Cambridge. He was my scientific father in every sense of the word. Rutherford didn't like his associates keeping long hours in the lab. He thought it was silly to overdo it. But this didn't mean that Rutherford stopped working at any time. One day we went home without having understood the results of an experiment and our telephone rang in the night at 3 a.m. My wife told me that the Professor wanted to speak to me. Rutherford said, 'I've got it. Those short-range particles are helium of mass three.' I asked him about his reasons and he said, 'Reasons! Reasons! I feel it in my water!'

Was he an approachable person?

He was very approachable. He was always ready to talk to anybody about anything under the sun; and he was also a tremendous influence that way. His booming voice could always be heard at the high table in college. The Rutherfords were very good in a social way, too. They had Sunday afternoon tea parties to which they invited the students. They had a cottage in the country to which they invited students to spend the weekend. It was always a great experience.[2]

Is it true that Rutherford did not believe that nuclear power was possible?

He hated the idea. He knew it was possible but didn't want it to be possible.[3]...

Once you gave a lecture entitled A Physicist's View of God.

I think the theoretical physicist is the closest to God, the theoretical physicist who thinks about the fundamental realities.

Do you think we should be looking for a unifying principle in nature?

I don't think so. I think all knowledge is important, and it's just as well that we have a very wide and chaotic approach to knowledge, that people of all sorts are seeking fresh information in every corner. That's as it should be. Nature is very complicated. The more one knows it, the more complicated it becomes. If it's the other way around, then one isn't thinking."[4]

---

[2] Oliphant wrote a book about Rutherford: *Rutherford: Recollections of the Cambridge Days* (Amsterdam: Elsevier, 1972).

[3] *Candid Science IV*, 307–9.

[4] Ibid., 311–12.

WOLFGANG K. H. PANOFSKY

The problem scientists have created is that fewer and fewer people can do harm to more and more people.

Wolfgang K. H. Panofsky in 2004 in Palo Alto, California. (photo by M. Hargittai)

Wolfgang K. H. Panofsky[1] (1919–2007) was born in Berlin, Germany, and immigrated with his family to the United States upon the Nazi takeover. He received his PhD degree in physics in 1942 from the California Institute of Technology. He was at Stanford University from 1951 to the end of his life. He organized and oversaw the creation of the Stanford Linear Accelerator Center (SLAC) in Palo Alto, California, and served as its director between 1961 and 1984. He received many awards and other distinctions, was an adviser to the US government, and was active in discussing arms control and disarmament. We recorded our conversation in 2004 in Panofsky's office at SLAC.

---

[1] I. Hargittai and M. Hargittai, *Candid Science VI*, 600–629.

Speaking about the danger of nuclear smuggling, how small can a hydrogen bomb be nowadays?

"They are not that small. There was one hydrogen bomb that could be handled by one soldier but it still weighed about 150 pounds. They are not that small but they are small enough so that you can stick them in a crate or put them on a truck. Carrying it in a suitcase you ought to be awfully strong and it would set off all sorts of alarms. But it is a very real problem. Senator Nun said that the limit of damage a terrorist can do is limited by the tool at his disposal. The reason why we have not used nuclear weapons is deterrence. Namely, had either the Russians or the Americans used nuclear weapons it would have been suicidal and that applies to all the nuclear weapons states. But when you talk about subnational groups, and when people believe that life in Heaven is better than life on Earth, then deterrence does not work. So there is a real risk there. The idea of detecting smuggling is a multilayered thing. You have to detect smuggling, but the most important thing is that you have to control the inventories at the source. Terrorists cannot make plutonium or uranium. They have to swipe it or buy it or bribe somebody."[2]

About the social responsibility of scientists.

"I feel tremendously that we have a responsibility and that's why I write poison-pen letters. But I myself went through a tremendous evolution. Right after Sputnik, Eisenhower created his presidential science advisory committee. I was a member of it from 1959 to 1964. I knew Eisenhower quite well. He met with the scientists all the time and we had really open debates and open committees. What was important is that the president was science literate; of course, he was not a scientist, but he calibrated our strengths and weaknesses pretty well. But that ability of the top administrators has eroded steadily. Kennedy understood it more or less. I met with Carter several times and he, in a funny way, overestimated what technical people could do. For example, I wrote a chapter on nuclear reactor safety. We had a one-hour appointment with the president, where he briefed me, and at the end, when we had only two minutes to go, he told me: 'Dr. Panofsky, can you explain to me the difference between the sodium fuel cycle and the uranium fuel cycle?' He was the President of the United States and surely he had strengths and weaknesses, but he was a curious man—but of course, no human being could explain in detail these things in two minutes. So I sort of chickened out. He listened to too much detail in science so he could not see the forest for the trees—but at least, he listened. But since then... .

How about Clinton?

Clinton was not that good. I was not directly involved with the president's council on science and technology but knew several people who were. They mainly talked to [Vice President] Gore and not to Clinton. Gore, again, understood science but he thought he knew the answers, so was not a very good listener. Clinton understood the value of science and he was technically literate but he did not take the time to really

---

[2] Ibid., 618–20.

understand some things. He was better though than the current [George W. Bush] administration. But it was Eisenhower who was very good in appreciating science. Of course, he was a university president so he knew the academic animal fairly well. Johnson had essentially no communication. It's not a matter of being a Democrat or a Republican; it's very much on the individual style and character of the particular gentleman. But anyway, that's the way it is."[3]

---

[3] Ibid., 628–29.

ROGER PENROSE

What interests me most is what's true, what is there out there, and to try to understand.

Roger Penrose in 2000 at the University of Oxford, UK. (photo by I. Hargittai)

Roger Penrose[1] (1931–) was born in Colchester, Essex, England. He received a bachelor of science degree from University College London and a PhD in algebraic geometry from Cambridge University. He is a mathematical physicist and cosmologist, but it is difficult to categorize him; he is also a critically acclaimed science writer. We recorded our conversation in 2000 in his office at Oxford University.

We asked Penrose whether he considered himself a maverick.

"Some people think so. They get upset by my work related to consciousness.... There is a view that either you believe that we are just computers or you are mystical or religious or you have some view which is regarded as unscientific.

[1] B. Hargittai and I. Hargittai, *Candid Science V*, 36–55

Is there a contradiction between your being a realist who says that the world is there whether or not an observer is present and your being an antireductionist who says there must be something else in addition to physics and chemistry?

It's part of the same philosophy because I'm saying that conscious phenomena are real things. There is a real phenomenon there; it's part of the real world.

Can you explain it?

Someday, yes. People haven't got the explanation yet. I'm only interested in what's true.

Have you encountered the view that we may not understand quantum mechanics because it's part of divine reality?

Yes, I have. Of course, there is fuzziness in boundaries here. When I say I'm not religious, it means that I don't believe any religious doctrine that I have seen. It doesn't mean that I don't think there's something more than what is described by a purely reductionist view of the world. We have to discover what's going on. The word reductionist doesn't have a clear meaning. Sometimes it just means scientific, and I have nothing against that. But if it means that you can explain large things in terms of the behavior of small things, I don't necessarily believe that. There's a lot we don't understand about the world.

What's the next step in learning more about it?

Quantum mechanics is very important and we have to discover what's really going on there. We don't have the right view yet. We need a new physics, which we don't have. Consciousness is not independent of this question; it's just the wrong way around. There are people who say that quantum mechanics needs to be completed by having consciousness, but that I don't believe. What I believe is that consciousness is a physical phenomenon, it takes place in the world, and we don't understand the basis of it yet. There's this missing borderline between the small quantum physics and the large-scale classical physics. This missing ingredient is much more important than almost anyone would claim. You talk with physicists and they think it's a minor problem that we just need to understand quantum mechanics better. What I think we need is another theory. When we'll have that theory, it will only be a little step toward knowing what consciousness is. So you see what I mean by saying that I'm going the other way around. I'm not saying that consciousness is needed for quantum mechanics, what I'm saying is that quantum mechanics needs improvement to understand what consciousness is. And it's not even the whole story. You'll need to know more things too. There is a lot more we don't know in the way the world operates than most physicists would claim.

You asked me about being a maverick. When I think of how could quantum mechanics be relevant in brain processes, that is the question I picked up on the suggestion by Stewart Hamaroff. There are microtubules, little tubes in neurons and in almost all cells in the body, but in neurons they have a particular role. If my ideas are going to make any sense, you need something, which is beyond the level of neurons for large-scale activity in quantum mechanics. This is a very controversial idea. People say that if you're going to have large-scale quantum coherence at the temperature of the human body, this is ridiculous, they would say—you can only get these large-scale quantum activities at very low temperatures, like a superconductor. I think this is a very narrow view because there are lots of structures we don't know. Even high-temperature superconductors are at much lower than body temperature, and we don't know why they work. It's premature to make too strong claims of what can be going on there. But something must be going on, which is not explicable in ordinary classical terms.

What's going on in the brain? People say it's just some kind of computational process. That's why I'm a maverick because I don't think this is an explanation."[2]

---

[2] Ibid., 53–55.

ARNO A. PENZIAS

I had to get my astronomer's license.

Arno A. Penzias (on the right) and Robert W. Wilson in 2001 in Stockholm. (photo by I. Hargittai)

Arno A. Penzias[1] (1933–) was born in Munich, Germany. Being both a Jew and a Polish citizen was a double predicament. It was as if both Germany and Poland conspired to annihilate him and his family. Finally, he escaped from Germany when he was five years old and ended up in the United States as refugee. He received his bachelor of science degree from the City College of New York and his master's and PhD (1962) degrees from Columbia University. He began his research career at Bell Laboratories, and between 1967 and 1985, he was also associated with Princeton University on a part-time basis. He shared half of the 1978 Nobel Prize in Physics with Robert W. Wilson "for

[1] M. Hargittai and I. Hargittai, *Candid Science IV*, 272–85.

their discovery of cosmic microwave background radiation."[2] We recorded our conversation in 2001 in Stockholm.

Penzias and Wilson's discovery in 1965 made waves because it was considered direct evidence for the originally ridiculed Big Bang model of the origin of the universe. The best-known promoter of the Big Bang model, George Gamow, was still alive.

"Gamow was very happy, although this was not a direct proof of his theory. He used the following analogy. I lose a nickel and you find a nickel. I can't prove that it's my nickel except that I lost a nickel, just there. People loved it. Gamow was very gracious. [Ralph] Alpher, on the other hand, was very upset. He claimed that he had predicted the microwave background. I looked at his paper and told him that he did not predict an observable microwave phenomenon. He predicted the existence of a background energy. In a later article in *Physics Today*, for example, he and Bob Herman had, in fact, written that the cosmic background would be undetectable because it is masked by starlight and cosmic rays, both of which had similar energy densities.

How great was Gamow as a scientist?

Personally, I think that he was a better scientist than Galileo, at least as far as their respective contributions to cosmology are concerned. Galileo's "proof" of the motion of the Earth around the Sun was based on the tides. The Earth undergoes a compound motion from its orbital motion and its rotation, so the side away from the Sun moves faster than the side facing the Sun. He made an agreement with the Catholic Church not to write a popular pamphlet until he had incontrovertible scientific proof of the motion of the Earth. Galileo's proof was based on the tides; but for his theory to work he needed one tide a day with a 24-hour cycle. People had known for centuries that there are two tides a day and Kepler demonstrated their relation to the Moon's phases. Some people criticize Gamow for the rough approximations he used, but as I've tried to show with my Galileo example, he wasn't the only one to do so."[3]

The discovery of the cosmic microwave background radiation launched Penzias's career in radioastronomy, but it did not make him a radioastronomer overnight.

"After the discovery of the cosmic microwave background radiation, I wanted to continue to expand the scope of my research. In 1972, I joined Princeton as an unpaid associate. I supervised PhD students and did a range of experiments with them on a number of subjects. After a while, I started working on interstellar molecules. I concentrated more on isotopes than on chemical composition, because of my background in physics. Using our millimeter technology to measure rotational spectra, we changed the rules of the observational game. Before us, only a couple of odd lines were known, such as the inversion transition of ammonia or some weird states of water. There were very few ways of looking at molecules. Once you get to millimeter spectrum, where

---

[2] "The Nobel Prize in Physics 1978," *Nobelprize.org*. Nobel Media AB 2013, http://www.nobelprize.org/nobel_prizes/physics/laureates/1978/.
[3] *Candid Science IV*, 277–78.

everything has rotational spectra, and it becomes very easy. In the first days, we found a bunch of molecules, starting with carbon monoxide. Then we did some isotope work. Now, hundreds of people are doing similar work. My own interest, from a cosmological perspective, was to use this to study deuterium. I was the first to measure deuterium in interstellar space. Suppose, you are a farmer in Egypt; as your plough goes down a furrow, you find this trap door, one which leads obviously to a tomb. What do you do? You would do well to cover up the door, and you go to archeology school, because the first archeologist down the steps discovers the tomb. You have to be an archeologist to get credit for the discovery. Having only published a couple of papers at the time of our cosmic background discovery, I had to get my astronomer's license anyway. What gave me my astronomer's license was not so much that we did the background radiation, but that we discovered the interstellar molecules."[4]

---

[4] Ibid., 285.

JOHN C. POLKINGHORNE

There is no incongruity in being both a physicist and a priest—it is not like being a vegetarian butcher!

John C. Polkinghorne in 2000 in his home in Cambridge, UK. (photo by I. Hargittai)

John C. Polkinghorne[1] (1930–) was born in Weston-super-Mare, England. He is a theoretical physicist and an Anglican priest. He is also a prolific author. In 2002, he received the highly prestigious Templeton Prize, awarded for work that advances the "progress toward research or discoveries about spiritual realities." In physics, he worked in theoretical elementary particle physics, on the formal mathematical side. He received his PhD degree at Trinity College, Cambridge, in 1955 and subsequently taught at Cambridge. In 1979, he resigned his professorship and trained for the Anglican priesthood. He was ordained in 1982. Initially, he served outside academia, but eventually he returned to Cambridge, while continuing his service in the Anglican Church. We recorded our conversation in 2000 at King's College and in Polkinghorne's home in Cambridge.

[1] M. Hargittai and I. Hargittai, *Candid Science IV*, 478–95.

Polkinghorne said this about his career change.

"Quite early on, as a professional physicist, I felt that I wouldn't stay in the subject all my working life. The reason was quite simply this: although I don't think that you do your best work before you are twenty-five in these mathematically oriented subjects; yet most of us do our best work before we're forty-five. I had seen a number of senior friends who had been very active in the subject, who as they moved into their late forties began to lose touch with physics as the subject moved away from them. They remained in senior, quite important positions, but they seemed to me somewhat miserable in those positions. They knew they weren't quite at the cutting edge in the way they had been before, and that's an uncomfortable situation to be in. If you have a responsibility for a lot of quite young people, who are working in your group, to give them proper leadership, you don't want to be too far from the action, although you don't have to be the very best researcher in the group.

So I'd long thought that I would not stay in physics all my life, and this feeling was reinforced by the progress in theoretical particle physics, where I was working, and which was changing very rapidly. I had lived through a very interesting time in the subject, essentially the whole period of discovering the so-called Standard Model, in which we, collectively, as a community, discovered that protons and neutrons are made of more-fundamental particles, the quarks and the gluons. It took about twenty-five years to figure all that out. It was a development that was largely experimentally driven, with the theorists limping along behind the experimentalists, continually adjusting their ideas to make sense of what was being discovered. It was a very interesting period to live through, but by the middle seventies, the dust had settled. Once the Standard Model had been established, the subject started to change character. It became more mathematical than it had been, more speculative than it had been, and more difficult than it had been. All these things, combined with my age, encouraged me to think that I had done my little bit for physics. I wanted to stay in touch with the subject, but no longer in a technically proficient way. That's when I left physics, but I did not act so because I was in any way disillusioned with the subject. I resigned my chair in Cambridge in 1979 at the age of almost 49."[2]

About being a priest and a physicist.

"I see no contradiction between the two; they are complementary to each other rather than in conflict with each other. They represent different perspectives on reality. Science and religion have different ways of investigating the things that interest them. What they have in common is seeking to respond to the way things are and desire to search for truth. There is no incongruity in being both a physicist and a priest—it is not like being a vegetarian butcher!

I think the two can be reconciled. I do a lot of thinking and writing and talking about how they can be held in mutual relationship with each other. It's quite a subtle business. It's rather easy to make crude statements which have a certain element of truth in them,

---

[2] Ibid., 483–84.

but which are too unsubtle to describe the situation fully. You can say that science is concerned with how things happen, meaning the processes, and religion is concerned with why things happen, meaning the purpose. However, the way you answer how questions and the way you answer why questions have to fit together. There has to be some consonance between the scientific view of the world and the theological view of the world. Seeking that consonance necessitates some exchange between them. For example, science can tell religion what the history of the universe has been and what its structure is, and religion has to listen to that. Science does not need outside help to answer its own questions. If you believe that the world is God's creation, which, of course, I do, and if you believe that the universe has had a long evolving history, starting very simple and becoming very complicated, that tells you something about how the Creator may be at work in that creation. God is patient and subtle, not in a hurry. God did not snap a divine finger to form things ready-made. In fact, in a very famous phrase coined immediately after the publication of *The Origin of Species*, God maintains a creation that's allowed to *make itself*. That's theologically how I understand an evolving world. There's the potentiality present in matter to produce life, to produce consciousness, and so on. That potentiality is explored by creation itself through the process of evolution. This way of thinking deepens and strengthens the theologian's understanding of what is going on in the world. What theology can do for science is not to tell science the answers to scientific questions but to take scientific insight and set in within a broader, more comprehensive matrix of understanding. For example, it's very striking to me that mathematics is the key to unlocking the secrets of the physical universe; that mathematical beauty is the actual tool for scientific discovery in fundamental physics. That seems to be a very strange thing, because mathematics seems to be such an abstract subject. Eugene Wigner called it 'the unreasonable effectiveness of mathematics,' which, he said, is something we neither deserve nor understand. Well, I'd like to understand it if I can, but it seems to me that science itself can't help me to do so. Science is simply very happy and content that mathematics has this marvelous power and then gets ahead with exploiting it. My religious belief makes me want to look at this issue in a more profound way. It seems to me that the mathematical beauty and intellectual transparency of the physical world shows the signs of mind. For me, that's consistent with my religious belief, that it is Mind, with a capital *M*, the Mind of the Creator that lies behind the marvelous order of the physical world. Here is the true source of that sense of wonder that scientists feel at the marvelous structure revealed to their inquiries. In that way, I experience a consonance between science and religion. They are different things, to be sure, but the question is, how do they fit together? Not without puzzles, of course, but I feel that I understand more if I look at the reality of the world in which we live in a two-eyed way, with the eye of science and with the eye of theology, than I would look with either eye on its own."[3]

---

[3] Ibid., 487–89.

DAVID E. PRITCHARD

Our strength is that [at MIT] we train outstanding young scientists.

David E. Pritchard in 2002 at MIT. (photo by I. Hargittai)

David E. Pritchard[1] (1941–) was born in New York City. He is Cecil and Ida Green Professor of Physics and principal investigator of the Atomic, Molecular, and Optical Physics Group at the Massachusetts Institute of Technology. He earned his bachelor of science degree at the California Institute of Technology and his PhD degree at Harvard University. He has been at MIT since 1968. He is a member of the US National Academy of Sciences and other learned societies. He has been a great mentor to younger scientists in addition to being at the frontier of research. He told us that he has mentored four future Nobel laureates. We recorded our conversation in 2002 in Pritchard's office at MIT.

We witnessed a moving moment during Wolfgang Ketterle's December 2001 Nobel lecture in Stockholm, when he asked Pritchard to stand up because he wanted to acknowledge Pritchard's contribution to the prize-winning work. When Ketterle was to become an

[1] M. Hargittai and I. Hargittai, *Candid Science IV*, 344–67.

Wolfgang Ketterle and his mentors at the Nobel Prize award ceremony in 2001 in Stockholm. From left to right: Herbert Walther, Wolfgang Ketterle, David E. Pritchard, and Jürgen Wolfrum. (courtesy of D. E. Pritchard)

assistant professor following his postdoc, Pritchard magnanimously gave Ketterle one of his own research areas. The project happened to be the one in which Ketterle made his Nobel Prize-winning discovery. Pritchard told us the little-known story of how he ended up with a Nobel Prize medal.

"As soon as we were both back from Stockholm, Wolfgang asked if I would like to see his medal (which I hadn't seen there). When I said yes, he quickly picked up his briefcase and led me into my office, putting several things on the table in my office. I thought this a bit strange because as a matter of courtesy, I should have stayed in his office to see his medal. But I chalked it up to Wolfgang's usual energetic approach to things.

To my surprise, he put three medals on the table. "Dave, they actually gave each of us three medals—the real gold medal, and two replica bronze medals plated with the same karat gold as used on the solid gold medal. Can you tell which is the real one?" I often tell the story of how Archimedes developed his principle for determining nondestructively whether the King's new crown was solid gold. His method relies on the fact that gold is the densest metal commonly available, being about 50 percent more dense than lead. So I was quickly able to pick out the gold medal even though the bronze ones were slightly thicker so that their weight was not that much less.

I had the gold medal in the palm of my hand and passed it to Wolfgang so that he could verify that I had correctly identified the gold one. Instead of taking it, however, he wrapped his hand over mine, closing my fingers over the medal, and said, 'You keep holding on to it—it's yours. I'm giving you this one. I'll give one of the replicas to the department and keep the other for myself.' My mother will be glad to hear that I remembered to thank him for this. But I was too stunned to express my true feelings to him until the next day.

In the meantime, we decided that it was appropriate for me to carry the medal in my computer case, as it is no more valuable than the computer, at least monetarily. Later, I mused on how Wolfgang's method of presentation had achieved three objectives at the same time—another example of his unmatched ability to find solutions that solve several problems at once. In case I had known about the replicas, it made sure I knew I was getting the gold medal. It also made me realize that he was not giving up his only Nobel medal. Finally, by getting it in my hand and holding the hand shut, he made it more difficult for me to decline the gift.

The gift of this medal has had two tremendously beneficial side effects. The medal has been enriched by being given. Furthermore, it is possible for me to show it and talk about its doubly positive reflection of Wolfgang (i.e., both winning it and giving it) without it seeming to be in bad taste as it would if Wolfgang himself were to exhibit it. I have passed it around at several department gatherings and at a gathering of the alumni of my research group. Many of those who hold it are deeply touched and thank me profusely for risking it being passed around. I doubt that any other Nobel medal has generated so much happiness in so many people.

That's wonderful, and such things obviously don't happen often. But I suppose giving exciting experiments away is fairly unusual also.

Well there is lots of precedent for giving away experiments in our group. Dan Kleppner led the way when he gave Bill Phillips the machine he'd constructed to do his thesis in—it housed the first successful atom slowing experiments at NIST [National Institute of Standards and Technology]. Then I gave my thesis machine to Stuart Novick at Wesleyan University, and later gave my PhD student Brian Stewart my experiment on molecular energy transfer and the large double monochrometer that went with it; he has produced new results with it for fifteen years. This year I'm giving my atom interferometer to my current postdoc, Alex Cronin; and next year I plan to give the mass spectrometer to Ed Myers at Florida State. I find it easier to stop doing old things and move on to new things if I know that my still-state-of-the-art experiments will go forward—often faster due to the infusion of new blood."[2]

---

[2] Ibid., 346–48.

NORMAN F. RAMSEY

The universe seems to run without an external god-like intervention.

Norman F. Ramsey, Douglas D. Osheroff, Charles H. Townes, and Magdolna Hargittai in 2005 in Lindau, Germany. (photo by I. Hargittai)

Norman F. Ramsey[1] (1915–2011) was born in Washington, DC, and studied at Columbia University and at Cambridge University (UK). At Columbia, he did his doctoral studies on magnetic resonance under I. I. Rabi's supervision. During World War II, Ramsey did defense-related work, and then spent his career, starting in 1947, at Harvard University. He was awarded the 1989 Nobel Prize in Physics "for the invention of the separated oscillatory fields method and its use in the hydrogen maser and other atomic clocks."[2] We recorded our conversation in 2002 in Ramsey's office at Harvard.

---

[1] M. Hargittai and I. Hargittai, *Candid Science IV*, 316–43.
[2] "The Nobel Prize in Physics 1989," *Nobelprize.org*. Nobel Media AB 2013, http://www.nobelprize.org/nobel_prizes/physics/laureates/1989/.

When we asked him about the greatest challenge in his life, he described his unsuccessful attempt to demonstrate violation of parity.

"Probably a major unsuccessful challenge was the search for parity nonconservation. Purcell and I had the first idea that parity might not be conserved, and we did an experiment on parity, testing it at a time when everybody thought that this is something one should not waste time on. This was about five to six years prior to the Lee and Yang paper. In fact, the only paper Lee and Yang quoted was one of our papers, even if they slightly misquoted it by confusing it with our later experimental limit, whereas the paper they referred to was the first one, which pointed out that parity symmetry was not obvious and should be tested experimentally.[3] We did an experiment in which we were looking for an electric dipole moment as a test of parity when everybody believed that it was a somewhat stupid experiment because parity had to be conserved. We worked on that for about five or six years, and did a very good experiment, but the problem was that we were looking at the nuclear strong forces. At that time people did not differentiate much between strong and weak forces. We did a very, very, sensitive experiment, but what we tested was the nuclear strong forces, and there, as far as we still know, parity is not violated. When Yang first reported on their theoretical work at an MIT colloquium, I immediately tried to arrange an experiment to do a parity search with weak forces, and arranged with Louis Roberts at Oak Ridge Laboratory, the only person who has ever aligned a nucleus in large amounts, which he did with Co-60. Unfortunately, soon after we made our arrangements, the Oak Ridge management postponed our experiment.

Was your experiment the same as the one Madame Wu did?

Yes, it was. I corresponded with Yang about this but I did not know then that Mrs. Wu was even considering an experiment.

What gave you the idea originally to look for parity violation?

That is an interesting question. Shortly after I came to Harvard, I was giving a graduate course on molecular beams. Ed Purcell, the co-inventor of NMR [nuclear magnetic resonance], who was a professor at Harvard that time, sat in the course because he was interested in the topic.

I discovered the following: it was fun having Purcell in the class because it led us to have interesting discussions. But I also learned one hazard of having Purcell in the class; if I was talking about a subject that I did not really fully understand, I could count on Ed asking an astute question that not only would convince me that I did not understand but would also convince him and the whole class.

In the class, I was about to give the well-known proof: if parity is conserved, there cannot be an electric dipole moment. All one knows about the orientation of the nucleus is its angular momentum, as if something is spinning. If it is spinning in one direction, the only way you can tell what's up and what's down is to grab it with your right hand and if your fingers go in the direction of the spinning, then your thumb

---

[3] E. M. Purcell and N. F. Ramsey, "On the Possibility of Electric Dipole Moments for Elementary Particles and Nuclei." *Phys. Rev.* 78 (1950): 807.

points in the up direction. If you do it with your left hand, you get the opposite answer. Therefore, if you have an electric dipole, that is a violation of parity. I knew this, and I knew how to give its proof. But then, I had the vision that Purcell would ask 'but what's the evidence for nuclear forces of parity being a good assumption?' so I thought, I better figure it out before he asks. I looked at all the experiments and I could not find any evidence. At that time, as I said before, we did not talk much about strong and weak forces, only simply nuclear forces.

I thought that the best way to defend myself against a Purcell attack would be a counterattack, so a couple of days before class I went to Purcell and told him about this. He said, oh, there must be lots of evidence. So he went to look for evidence, and he could not find any either. So eventually, we wrote up a paper saying that we would do an experiment in which we would look for electric dipole moment as a test for parity."[4]

---

[4] E. M. Purcell and N. F. Ramsey, "On the Possibility of Electric Dipole Moments for Elementary Particles and Nuclei." *Phys. Rev.* 78 (1950): 807; M. Hargittai and I. Hargittai, *Candid Science IV*, 340–41.

VERA C. RUBIN

If the community of scientists was more welcoming, many more women would succeed.

Vera C. Rubin in 2000 in her office at the Carnegie Institution in Washington, DC.
(photo by M. Hargittai)

Vera C. Rubin[1] (1928–), astronomer, was born in Philadelphia. She received her bachelor of arts degree at Vassar College in 1948, her master's degree at Cornell University in 1951, and her PhD degree at Georgetown University in 1954. George Gamow was her mentor in her doctoral studies. She is an associate of the Department of Terrestrial Magnetism of the Carnegie Institution in Washington, DC. Her most famous discovery was that most of our universe consists of dark matter. Some of her discoveries concerning galaxies were ahead of their time, but eventually the scientific community recognized their validity. We recorded our conversation in 2000 at the Carnegie Institution.

---

[1] B. Hargittai and I. Hargittai, *Candid Science V*, 246–65.

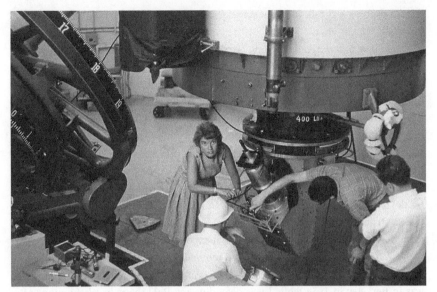

Vera C. Rubin (center) in 1965 at Lowell Observatory in Flagstaff, Arizona, with the 60-inch telescope. (photo by Bob Rubin; courtesy of V. C. Rubin)

Since Rubin had been George Gamow's graduate student, we were eager to ask her about him.

"Gamow had heard of my master's research. When we moved to Washington, my husband went to work at the Johns Hopkins Applied Physics Lab (JPL) and worked with Ralph Alpher and Bob Herman. Ralph, who had earlier written his thesis under Gamow, and Bob Herman were still working with Gamow on problems of the early universe. Through this connection, Gamow called me on occasion to discuss galaxies. Ultimately, his question, is there a scale length in the distribution of galaxies? seemed to be a good problem for a PhD thesis. Even though I was enrolled at Georgetown University (the only local college which offered a degree in astronomy) and Gamow was a professor at George Washington University, I made arrangements to write my thesis under his direction.

What was it like working with George Gamow?

He was very pleasant, very amusing, liked scientific games and jokes, but he was not really interested in the scientific details of an analysis. Except for very general guidance, he was not involved in the details of my calculations, but he was enormously interested in the implications of the results: patterns exist in galaxy distribution, and galaxies are strongly clustered.

How did he take it that his initial suggestion for the origin of the universe was not well received?

He was not disturbed. He enjoyed throwing out many ideas, important and imaginative ideas, about the universe. And he had great delight in those that survived. He was very generous, and liked the interactions and the social aspects of science. He started almost every talk with amusing scientific demonstrations. He liked to have fun, especially with science.

Could it be that the poor reception prompted him to turn his interests toward molecular biology?

I do not think that this had anything at all to do with it. Gamow was a physicist, but interested in, and curious about, all of science. He understood the important implications of what was about to happen in biology, and he was versatile enough to have important ideas in this field too. He went wherever his curiosity led. I think I am correct in stating that he was the first to devise a helix (but a single, not a double helix) for the DNA structure, noting that the four nucleotides would pair to form twenty amino acids. He made an early model of the DNA helix, formed a RNA tie club of twenty (male) leaders in the then baby field of DNA structure, sent each a tie with one of the twenty structures, ties and tie pins which he had specially made. I think he understood from the start that the DNA was a code that expressed the 'language' of biology.

How did he react to the discovery of the remnant heat by Penzias and Wilson?

Gamow left Washington in 1956, two years after I had earned my PhD degree. He was in Berkeley in 1954 while I was defending my PhD thesis in Washington. After that, I saw him at meetings or occasional visits before his death in 1968. I had only a few discussions with him about many of these matters. The last time I spent any time with him was at a meeting on relativistic astrophysics (the 'Texas' meeting that was that year held in New York). I am certain that he would have been delighted with the discovery of the remnant radiation by Penzias and Wilson. It proved that the cosmology of Alpher, Herman, and Gamow was very close to the cosmology inferred from the Penzias and Wilson observations.

How do you feel about the fact that Penzias and Wilson received the Nobel Prize and Gamow never received any formal recognition?

How do *I* feel? Well, let me say the following: my husband Bob and I remained very close friends with Bob and Helen Herman. We continue to see Helen now, some years after Bob's death. I know that he and Ralph Alpher believed that their work was never properly recognized, and were sometimes very bitter about this. I feel sorry for scientists who believe that their work was not given due credit, but I don't think that Gamow thought that his work had been neglected. It is recognized that important ideas in nuclear physics came from Gamow. His joy came from interactions with scientists, with dreaming up ideas that might be correct. He had a distinguished career, and he was recognized for his many innovative ideas. He did get recognition. He was an active member who enjoyed his activities with the US National Academy of Sciences."[2]

[2] Ibid., 249–52.

DAN SHECHTMAN

Linus Pauling is wrong.

Dan Shechtman in 1995 in Balatonfüred, Hungary. (photo by I. Hargittai)

Dan Shechtman[1] (1941–) was born in Tel Aviv. He is Distinguished Philip Tobias Professor of Materials Science at the Technion—the Israel Institute of Technology in Haifa. He is best known for his 1983 discovery of quasicrystals, for which he was awarded the Nobel Prize in Chemistry in 2011. The quasicrystals are ordered but nonperiodic structures whose existence was considered impossible according to the rules of classical crystallography. Thus Shechtman's discovery brought about a paradigm change in what we consider to be a crystal. Linus Pauling, one of the greatest chemists of the twentieth century, opposed the interpretation of Shechtman's experimental observation. This giant's opposition first hindered the recognition of Shechtman's achievement, but in the final account, its challenge probably stimulated the rigorous understanding of the discovery. We recorded our conversation in 1995 in Balatonfüred, Hungary, on the occasion of an international school on quasicrystals.

---

[1] B. Hargittai and I. Hargittai, *Candid Science V*, 76–93. See also chapter 8 in I. Hargittai, *Drive and Curiosity* (Amherst, NY: Prometheus, 2011).

The Hargittais and the Shechtmans in December 2011 in Stockholm. (photographer unknown)

We reminded Shechtman that in what may have been Pauling's last interview, he was still insisting that he did not believe in quasicrystals. Shechtman told us about his interactions with Pauling.

"Linus Pauling heard about the discovery and contacted me in writing. He wanted some information, which I sent him. Then he wanted more information, and he sort of complained that I was not correct in the information that I had provided him with, which I was. Nevertheless, I repeated the work, performed the needed microscopy again, and sent him a short paper that I wrote just for him. He replied by saying that it was OK, and what I did was fine, except he did not agree with the interpretation.

Then at a certain stage I suggested to him that I'd come and visit him in Palo Alto, and show him the result. I went and gave a full lecture to an audience of one. He had many questions, which I answered, but he was very negative, and he still did not believe in this. I showed him results, which for me were very conclusive. He said, 'I don't know how you do that.' If it were a student, I would probably say 'OK, go and read a book if you don't know how to do that'. But with Linus Pauling? This is the man who wrote the books. Anyway, as I was leaving, I asked him, 'If you change your mind, and if you ever agree with me, please publicize it, and let it be known.'

Then we met several more times at conferences. It was a friendly meeting every time, and we invited each other to dinner. People were looking at us as if expecting a fistfight, but the conversation was always very pleasant. We also agreed on many things, like the importance of Vitamin C, but never on quasicrystals. A couple of years later, I was attending a big lecture by him at Stanford, organized by the American Chemical Society.

The topic of quasicrystals came up in his talk, and he mentioned how bad it was. I was just sitting in the audience and nobody knew me. He was like a mixture of a politician and a priest. He had this quality to become a charismatic leader, enjoying the admiration of the crowd, with no questions asked. In his effort to explain the icosahedral phase as periodic, he presented a model of twinned crystal, which was very soon afterward proven wrong by others. However, in this lecture he had the floor, of course. He talked about me by name, but he didn't know that I was there. At a certain point, I turned to the man sitting next to me and said, 'Wow, he's wrong.' And he said, 'What?' and I repeated, 'Linus Pauling is wrong.' And he shouted, '*What*???' as if he was going to hit me. It was a fanatic crowd.

In 1987, I met several of Pauling's close disciples in China. They came to the second international conference on quasicrystals. Each one of them separately and as a group said, Danny, we know you are right. And I said, hey, this is very important; I need this in writing. But then they said, we can never put it in writing because it would kill our Linus; he trusts us, and we can't betray our old master. I felt very bad about it; I felt that science should not be done this way.

At one time Linus wrote me a letter suggesting that we publish a paper together and settle the differences between us. I answered him with a letter saying the following: I'll be honored to write a paper with you, but we have to agree on the principles first, the first principle being that quasicrystals exist, and they are not twinned crystals. He wrote me back and said that maybe it was too early to do this joint paper."[2]

---

[2] *Candid Science V*, 91–93.

VALENTINE L. TELEGDI

I don't like to shoot with cannons at sparrows; some people consider this a form of elegance.

Valentine L. Telegdi and Mrs. Telegdi in 2004 in Pasadena, California. (photo by M. Hargittai)

Valentine L. Telegdi[1] (1922–2006) was a Hungarian-born American experimental physicist who had an adventurous youth between the wars and during World War II. He received his PhD degree in 1950 at the Swiss Federal Institute of Technology. He was at the University of Chicago between 1951 and 1976 and returned to the Institute between 1976 and 1990. In retirement, he divided his time between CERN (the European Organization of Nuclear Research) in Geneva and Caltech in Pasadena, California. He was an interesting and outspoken individual. At the age of forty, he formulated his goal in

---

[1] M. Hargittai and I. Hargittai, *Candid Science IV*, 160–91.

Valentine L. Telegdi and Wolfgang Pauli in 1957 during the Padua-Venezia Conference. (courtesy of V. L. Telegdi)

life—to be respected by those people whom he respected.[2] He fully achieved this goal. Telegdi, together with Jerome Friedman, performed one of three famous experiments in 1956 and 1957 that proved the nonconservation of parity in weak interactions. We recorded our conversation in 2002 in our home in Budapest.

We remarked that the Soviet Sputnik had become an important driving force for American science.

"But not only for science, for the general public, too. In fact, more for the public than for science. It also had a positive effect in helping science, but not as deep an effect as some people think. People hoped that it would do a lot for improving American education. Not so much American research but American education. People realized that a certain type of schooling was much better in Russia than in the United States, and they wanted to do something about that.

The support of science in the US had much to do with the war and Los Alamos. It was the first time that the military had understood how valuable science could be to the country. That is the origin of the large-scale support of science. The sum of money that scientists spent during the war years was still very small compared to military expenditures. There was a certain feeling of gratitude towards the scientists. Unfortunately, all that is gone now, all those people who were grateful are by now dead. So we have to start all over again. In America many people confuse science with technical progress. They think that science will produce faster cars and better toothbrushes, but that has very little to do with science."

---

[2] Ibid., 190.

On why Einstein did not like quantum mechanics.

"Because he was putting philosophical considerations ahead of physical considerations. He included a certain type of philosophy, which extended to Nature. He had no real prejudice about physical laws; in fact he could do the wonderful things that he had done because he could think completely freshly about any subject. But he had deep philosophical principles. He deeply believed in determinism. He did not say that quantum mechanics was wrong, he just said it was not satisfactory; that he could not accept a theory that makes only probabilistic predictions. This is really an expression not of his thinking of physics but about his own philosophy. Why not? There are many puzzles at the foundations of quantum mechanics that people keep discussing and there are many people still discussing these difficulties. There is a thing which I call 'Alka Seltzer physics.' What is Alka Seltzer physics? A man tells me 'when I think about such and such a proposition in quantum mechanics, I feel sort of ill in my stomach.' Well, take Alka Seltzer. I cannot judge physics on the basis of anybody else's emotions. It is very individualistic what bothers one's emotions and what does not. It is not when we say that something contradicts an experiment. Determinism is a way of looking at the world. I am not thinking about these things nowadays. There was a time when I used to discuss them with Gell-Mann at great lengths but that was more than ten years ago.

Fermi had a very simple point of view: Quantum mechanics is correct because it works. Here is a machinery that gives results and agrees with Nature, so he never discussed these problems, they never bothered him at all. He was completely pragmatic and he was a very great man"[3]

---

[3] *Candid Science IV*, 180.

EDWARD TELLER

I am an anticommunist essentially of the school of [Arthur] Koestler.

Edward Teller and Istvan Hargittai in 1996 in Stanford, California. (photo by M. Hargittai)

Edward Teller[1] (1908–2003) was born in Budapest, Hungary. He made major contributions to physical chemistry and to atomic, nuclear, and solid state physics. He was also involved in the Manhattan Project, the development of the hydrogen bomb, and the Strategic Defense Initiative. He initiated and was the leader of the Livermore Laboratory, and fought against the banning of nuclear-weapons testing. He testified against J. Robert Oppenheimer in the 1954 hearings about Oppenheimer's security clearance. Teller was a controversial figure, and many have found it difficult to form a balanced opinion

---

[1] M. Hargittai and I. Hargittai, *Candid Science IV*, 404–23.

about him.[2] Teller started his education in chemical engineering, but eventually became a world-famous physicist. He was a recipient of the Enrico Fermi Award and the Albert Einstein Award, among others. We recorded our conversation in 1996 in Stanford, California.

We asked Teller whether he had seen the criticism of him by some people who maintained that his opinion was colored by his history back in Hungary.

"Often. Let me tell you, when Hungary went communist in 1919, I was eleven years old. The communists were in Hungary for four months. Our family didn't like them at all. My father, as an advocate lawyer, had no job. Then there was the counterrevolution, and there were shootings in the streets, which I, from quite a bit of a distance, witnessed. I was certainly anticommunist, but at the age of eleven, I was not exposed to the problems of communism in a direct way. I was exposed in the school to anti-Semitism.

In 1926, I went to Germany to study. This was just two weeks before I turned eighteen, after I'd completed Gymnasium and attended the Budapest Technical University for a few weeks. I just finished the first term there in chemical engineering. My father did not approve of mathematics, and we compromised on chemical engineering.

At that time, I can honestly say that I had an open mind about communism. In 1926, the horrors of the Stalin regime had not yet become obvious. In Europe, at that time, communism was considered the way of the future, and particularly so after 1928–29, after the big, worldwide depression that was clearly the end of capitalism. The only people who had new ideas, so it was said, were the communists. Now, I was not a communist, not at all, but I was genuinely open-minded. I can best illustrate this to you by telling you about my friends. After studying for two years in Karlsruhe, I went to Leipzig to work with [Werner] Heisenberg. There I was exposed to a small, interactive, international community. Politics was not in the center of interest, but I had two very close friends—one, a clear anticommunist, the other, a devoted communist. The anticommunist was Carl Friedrich von Weizsäcker, the elder brother of the man who later became the president of the West German Republic. He was an excellent physicist, later a philosopher, and a close friend, and we studied together in Leipzig, in Göttingen, and later in Copenhagen, where we lived in the same house. I disagreed with him about communism. He was clearly anticommunist, and he was at that time also anti-Nazi, but not as clearly. Of the two evils, he considered Nazism the lesser evil.

Another good friend at that time, with whom I also published a paper, was Lev Landau. He was a devoted communist. He considered capitalism as something ridiculously wrong, and talked about it a lot.

I want to mention to you another friend, a Hungarian, Laszlo Tisza.... Tisza was very strongly pro-communist, not a member of the party but he worked with them. When he went back to Hungary, he was arrested. I continued to work with him and

---

[2] One of us produced what he hoped was a balanced biography: I. Hargittai, *Judging Edward Teller: A Closer Look at One of the Most Influential Scientists of the Twentieth Century* (Amherst, NY: Prometheus, 2010).

visited him in prison in Budapest. Eventually he was let go but his chances for a job were precisely zero. I then wrote to Lev Landau who got a job for him in Kharkov. In a few years he came back from the Soviet Union. He was completely disillusioned and told me about the horrible treatment of Landau, who had been arrested. This was about the time I came to the United States.

I have been an anticommunist, but not on the basis of what I had experienced in Hungary. It was after the conversations with Tisza that I just described, when he returned from the Soviet Union. This was in 1936. I can also tell you that the capstone of all this was a book, of which you know.

Darkness at Noon *by Arthur Koestler?*

Darkness at Noon. I read *Darkness at Noon* in the first month I was in Los Alamos. That was in the spring of 1943. The book could not have been published much before then. By that time I did not need much convincing. It is an excellent book because throughout the book Koestler holds you in suspense of which side is right. Then he comes out very strongly against Stalin in the last 50 or 100 pages.... So the statement that I am an inveterate Hungarian anticommunist is just not so. I am an anticommunist essentially of the school of Koestler."[3]

---

[3] *Candid Science IV*, 407–9.

CHARLES H. TOWNES

[The laser] has been a very useful scientific tool. You might even compare it to the screwdriver.

Charles H. Townes in 2004 at Berkeley. (photo by I. Hargittai)

Charles H. Townes[1] (1915–) was born in Greenville, South Carolina. He graduated from Furman University in 1935, received a master's degree in physics from Duke University and his PhD degree in physics from the California Institute of Technology in 1939. He worked for Bell Labs from 1933 to 1947 and served on the faculty of Columbia University between 1948 and 1961. Between 1961 and 1967, he was at the Massachusetts Institute of Technology, and since 1967, he has been at the University of California, Berkeley. He was co-recipient of the 1964 Nobel Prize in Physics together with Nikolai G. Basov and Aleksandr M. Prokhorov 'for fundamental work in the field of quantum electronics, which has led to the construction of oscillators and amplifiers

---

[1] B. Hargittai and I. Hargittai, *Candid Science V*, 94–137. This excerpt is taken from two interviews; one was conducted by Clarence Larson in 1984 (96–119); and the other by one of us [IH] in 2004 (119–37).

The young Townes with his experimental setup. (courtesy of C. H. Townes)

based on the maser-laser principle."[2] We recorded our conversation in his Berkeley office in 2004. One year after our conversation, Townes was awarded the Templeton Prize, which honors a living person who has made an exceptional contribution to affirming life's spiritual dimension, whether through insight, discovery, or practical works.

Townes told us about the separation of religion and laboratory work.

"I am religiously oriented and I think that many more scientists are religiously oriented than the public recognizes. They don't talk about it because when they talk about it, they get criticized, but more and more are becoming more open. The discussions about the interactions between religion and science become more public and I've given lectures on that subject. My own point of view is that religion and science are really very similar, much more similar than people recognize. Science involves assumptions or we can say faith. One of the basic assumptions in science is that this universe is reliable and is controlled by fixed laws and these laws are reliable and we can trust them and

---

[2] "The Nobel Prize in Physics 1964," *Nobelprize.org*. Nobel Media AB 2013, http://www.nobelprize.org/nobel_prizes/physics/laureates/1964/.

so on. That's an assumption. We don't know it for certain and the whole thing might change tomorrow. We can't prove it wouldn't, but we have the faith that it doesn't. Ingrained in science is that the same is true every day. That's one extreme case of faith. Science and religion also both involve experiment, observations. We observe people, history, how society works. Take astronomy, for example. We don't play with the stars up there, we look at them, we watch them, which is observation. Religion observes, and from that we try to conclude what is life about. You make observations and you make conclusions and you try to use logic.

Much of the discussion these days involves the question of was there a Creator? Many scientists for a long time believed that the universe never had a beginning and could not have a beginning. Einstein felt that and this is why he put in a cosmological constant because without this cosmological constant the stars would pull themselves together and the universe would collapse. He put in the cosmological constant to keep the stars apart because it had to be always the same. Then Hubble found that the universe was expanding and Einstein felt that he had made a mistake and threw away the cosmological constant. If something is expanding, it must have started from something smaller. Alternatively, some scientists have thought that new matter is being created all the time so instead of changing with expansion, it was always the same. Fred Hoyle pushed that a great deal and I talked with him about it. He gave lectures on this subject and I would point out to him that logically it could not be right. He admitted that it could not be right, but maintained that the universe always had to be the same. Just had to be the same. That was his faith. Even after the Big Bang was discovered, he continued this interpretation, but finally, he had to give it up.

Everybody recognizes now that if the Big Bang was not the beginning, at least it was a unique moment in the past. In addition, scientists are becoming more and more convinced that this is a unique universe. There have to be very special physical laws that allow everything to come out just right and us to exist. It has to be very special. If you are not religious you have to assume that everything just happened by chance and there may be many other universes with their own physical laws. If you want to say that no, this wasn't planned in any way, then there may be billions of other universes and ours just happened to turn out this way. The assumption of there being many other universes is also a question of faith because it cannot be tested. Religion fits the observation that what we have is very special and has come out exactly or almost exactly as it should have. These are the kinds of discussions that are going on more and more

My own assessment is that, yes, there seems to have been a plan; and I feel the presence of a God, or you might say a Spiritual Being if you don't want to call it God. There is something in the universe beyond what science usually talks about. There is no reason why science shouldn't take this up; and people could make tests. In fact, people do make tests, on the effectiveness of prayer, for example. We are making tests all the time because we are observing each other. We are seeing how people behave and trying to answer the question of what really makes a good life. That's my conclusion.

Don't you find mind boggling the question, if you are a believer, Where did the Creator come from?

It is mind boggling, of course. We can't visualize a beginning. In the beginning God created this universe, but who created God? How did it begin? The beginning is always a problem. There are great problems in religion and there are great problems in science. People don't recognize how many uncertainties there are in science. For example, quantum mechanics and general relativity are not consistent with each other. We have known this for a long time, but we believe in both of them. It is a problem. In addition, the zero-point fluctuations, which quantum mechanics predicts, produce an enormous amount of mass, more mass than all the universe or the energy of all the universe, but it isn't here somehow. Yet we believe quantum mechanics, though this is what it says. So there are a lot of inconsistencies within basic physics. Physicists are accustomed to these inconsistencies, and we just kind of push them aside. We see inconsistencies in what we understand about religion too. My point of view is that we have to do the best we can and accept the inconsistencies. Let's make the most logical and sensible conclusion that we can as to what the realities are. That's what I try to do."[3]

---

[3] *Candid Science V*, 133–35.

MARTINUS J. G. VELTMAN

Accelerator experiments are experiments where we have full control of all circumstances.

Martinus J. G. Veltman in 2001 in Bilthoven, Holland. (photo by M. Hargittai)

Martinus J. G. Veltman[1] (1931–) was born in Waalwijk, Holland. He received his PhD degree from the University of Utrecht in 1963. He started his career at CERN, and then worked at the University of Utrecht. Between 1981 and 1997, he was the John D. MacArthur Professor of Physics at the University of Michigan at Ann Arbor. After retiring in 1997, he returned to Holland, and lives in Bilthoven. He and his former student Gerard 't Hooft were jointly awarded the 1999 Nobel Prize in Physics "for elucidating the quantum structure of electroweak interactions in physics."[2] We recorded our conversation in 2001 at Veltman's home in Bilthoven.

---

[1] M. Hargittai and I. Hargittai, *Candid Science IV*, 80–109.
[2] "The Nobel Prize in Physics 1964," *Nobelprize.org*. Nobel Media AB 2013, http://www.nobel-prize.org/nobel_prizes/physics/laureates/1999/.

The achievement for which Veltman received his Nobel Prize represented the greatest challenge in his life.

"I have a hard time seeing it [from] that perspective. The greatest challenge is the things that you want to do. Maybe it was developing the theory of weak interactions that certainly was the biggest challenge; taking that direction was the crucial step, much of the other stuff was just a logical consequence of this. I knew the topic was important from the moment I started working on it, but the first moment when I saw I was right with my solution—that was a very peculiar thing, a very definite moment of my life.

Did you know that it was of Nobel Prize caliber?

Of course. But there was also an educational element in it that we have to appreciate here, because it is important. When I grew up in Holland I did not study with people who were in the front lines of particle physics. Even when I went to CERN—although CERN was the best in Europe—it was still quite secondary to the United States at that time. All the big shots were Americans: Feynman, Gell-Mann, Lee, Yang; they were all Americans, with no Europeans among them. A younger generation was starting, to be sure, but we were not aware of that at that time. When you get to that stage and you really want to do an important contribution, you have to be in an environment that puts you in that direction. I was at CERN, which was good, but was not good enough for that—it still is not good enough for that.

Then I went to SLAC [Stanford Linear Accelerator Facility] to America, and from that time I gradually grew up and I went back to CERN. I could not possibly have done the things that I did in 1968 earlier, in 1964, because there was still a considerable piece of education I had to go through. I had to learn what is important, to learn directions; otherwise you just do what everybody else does; a little aspect of a known theory. To start thinking in an original way, to find a new direction, to ask the real question, you have to learn to be fit for. This is something 't Hooft never had to do; of course he would never admit that because he never learned that.

I think I started to mature only in around 1966. Then I came to the level where it was possible to choose my direction, understand the complications, know the perspective, everything. I see this very often, when people come from places where they could not learn these things; they just don't see the perspectives, you even want to shout at them, but they would not listen. They are manipulating equations and they think that will solve something. For physics, mathematics is important, but the basic ideas come from somewhere else and once you understand where to go, it becomes a triviality. Then you use mathematics to solve the problem—but you can also get another guy to do that part for you. So there is that particular part of learning, which today, in this country, people don't do; they think that doing a lot of complicated mathematics is the right way to do physics. This is the big disadvantage of any place where there is not a first-class person to teach young people the sense of direction, the sense of relative importance, and the sense of relative unimportance of mathematics."[3]

---

[3] *Candid Science IV,* 105–7.

Veltman's opinion about whether or not astrophysics and particle physics are converging differs from that of many others.

"They are not converging. The astrophysicists use a lot of our knowledge, but we learn very little from them. Astrophysics is much less hard science than particle physics. They speak about the Big Bang, but much of it is basically speculation. Very little of it ever leads to better understanding of particle physics. It's a one-way street. Then, on top of it, astrophysics is not a very strong science."[4]

---

[4] Ibid., 108.

STEVEN L. WEINBERG

All good scientists have to rely on their intuition, on their taste of what an attractive theory would be, in the historical sense they feel this field is moving.

Steven L. Weinberg in 1998, at the University of Texas at Austin. (photo by I. Hargittai)

Steven L. Weinberg[1] (1933–) was born in New York City. He earned his undergraduate degree at Cornell University and his PhD degree from Princeton University in 1957. He is a theoretical physicist. He spent periods of time at Columbia University; the University of California, Berkeley; the Massachusetts Institute of Technology, and Harvard University, before moving to the University of Texas at Austin in 1982. He received the 1979 Nobel Prize in Physics together with Sheldon L. Glashow and Abdus Salam "for their contributions to the theory of the unified weak and electromagnetic interaction between elementary particles, including, inter alia, the prediction of the weak neutral current."[2] We recorded our conversation in 1998 at Weinberg's office at the University of Texas.

---

[1] M. Hargittai and I. Hargittai, *Candid Science IV*, 20–31.
[2] "The Nobel Prize in Physics 1979," *Nobelprize.org*. Nobel Media AB 2013, http://www.nobelprize.org/nobel_prizes/physics/laureates/1979/.

When we met with Weinberg he had just returned to his office from the lecture theater.

"I'm teaching a course on supersymmetry. It's a subject I last taught fifteen years ago here and at Harvard. Supersymmetry is a conjectured symmetry, for which there is no direct evidence yet. It would combine particles of different spin in the same symmetry multiplets. This is very different from the kinds of symmetries that we're familiar with, like the symmetries that underlie the theory that unifies the weak and the electromagnetic interactions, which only combine particles of the same spin, say, a neutrino and an electron.

There are various reasons to believe that nature, probably at some scale of energies, exhibits supersymmetry, although it is certainly a broken symmetry which is only manifest when you get to sufficiently high energy. There's a widespread feeling among particle physicists that the energy at which supersymmetry will be manifest is really just around the corner in today's experiments. Most likely the kind of experiment that will reveal supersymmetry will use the large hadron collider, which is a somewhat smaller counterpart of the supercollider. The supercollider won't be built, but the hadron collider will be built, early in the next decade. If supersymmetry is not discovered before then, I would bet that it will be discovered in the hadron collider. The discovery of supersymmetry is imminent and it's a good time to teach the subject."[3]

On the huge domain of nature in between what is covered by astrophysics and particle physics.

"In a fundamental sense we already understand nature on the level of the ordinary scale in which we live our everyday life. We have an understanding of the way atoms and molecules behave and the way electric forces and magnetic forces behave. This is sufficient to provide a basis for understanding the behavior of matter at ordinary scales. I'm not saying that we do understand the behavior of matter at ordinary scales; there are all kinds of wonderful things to be discovered in chemistry, in condensed matter physics, and elsewhere. We don't yet understand turbulence, and we don't understand intelligence. There are all sorts of things we don't understand. But we have a basis for understanding which doesn't require much more information. It's not likely that the behavior of chemicals or the behavior of living matter is going to reveal anything to us about the laws of nature at the most fundamental level.

This makes people angry when I say it, but by 'fundamental,' I don't mean interesting or exciting or valuable, useful, mathematically profound. It just means something that has to do with the rules that govern everything. By this I'm not saying that it's necessarily useful, for instance, for a chemist always to think in terms of solving the Schrödinger equation for a chemical molecule. Chemists have intuitions and expertise, which is often much more valuable than just the brute force solution of the Schrödinger equation."[4]

---

[3] *Candid Science IV*, 22.
[4] Ibid., 27–28.

On the personal element in science.

"I don't think that scientists behave objectively. I don't think they would get very far if they did— because we never know enough, so that you can't judge things just on the basis of some mechanical assessment of the experimental evidence. All good scientists have to rely on their intuition, on their taste of what an attractive theory would be, in the historical sense they feel this field is moving. That's very subjective. Then we all argue with each other, and that's a social process. We interact socially with other scientists in a complicated way. This is fascinating for a sociologist to study. But I think we converge on an objective truth which is what it is because that's the way the world is. In the end, it becomes a stable, permanent part of the body of our knowledge. At any one moment, the work of Kepler, Galileo, Copernicus, and Newton was highly influenced by the times they lived in, by the kind of people they had interactions with, and by their religion. But it all converged into a picture of the solar system we now have, and we have no doubt about it. Many of the sociologists who had studied science as just another social phenomenon, miss this. What they see is a process, and the process is like any other social process, the final result we are driven to is determined by nature, and we can't resist in the end knowing what the answer is."[5]

On the challenge of communicating with the "unwilling public."

"The best hope of science is to become part of the culture of our times, yet we are not, really, because the public can't read our articles. Even when we try very hard to write in a nonmathematical way, it's only a very limited segment of the public that would consider reading it. I have many dear friends who would never read one of the books that I've written for the general public. This is because anything scientific at all immediately loses their interest. I don't know what to do about it, we just have to try harder."[6]

---

[5] Ibid., 29.
[6] Ibid., 31.

JOHN A. WHEELER

I am proud that I participated in the Manhattan Project.

John A. Wheeler in 2001 at Princeton University. (photo by M. Hargittai)

John A. Wheeler[1] (1911–2008) graduated from Johns Hopkins University, and did his postdoctoral work with Gregory Breit at New York University and Niels Bohr in Copenhagen. From 1938 to the end of his life, he was associated with the physics department of Princeton University, with the exception of a ten-year period between 1976 and 1986 he spent at the University of Texas at Austin. He worked out the theory of nuclear fission together with Niels Bohr, participated in the Manhattan Project and the hydrogen bomb project, was Richard Feynman's mentor, and collaborated on investigations of the concept of action at a distance. He made seminal contributions in other areas of physics, and has been called "one of the most versatile physicists of the twentieth century."[2] Our originally published interview (see footnote 1) resulted from a series of conversations (with MH) in the early 2000s when our (MH and IH's)

[1] M. Hargittai and I. Hargittai, *Candid Science IV*, 424–39.
[2] Ibid., 426.

A group of participants of the theoretical physics conference at George Washington University in 1937. In front, Niels Bohr, second row, from left, Isidor Rabi and George Gamow, third row, Fritz Kalckar and John Wheeler, and fourth row, Gregory Breit. (courtesy of John Marlow, Princeton)

daughter, Eszter, was a doctoral student at Princeton University; hence our frequent visits there at the time.

We asked Wheeler two questions; the first was what he thought his most memorable achievement was.

"Richard Feynman regarded the one that when you shake a particle here, the reason that it loses energy is because the faraway matter of the universe interacts with it. I went to talk about that with Einstein. He was pleased with it. This brings together cosmology, the faraway, and radiation reaction, which is close up. Various people have applied that idea in other areas but nothing ever gave me that thrill. You see this arm and you see this leg? I would be glad to give them away, if in return, I could understand 'how come the quantum?' 'How come existence?' They do not seem to be the same question, but the answer will be the same."[3]

Our other question was about his most important achievement.

"The most important? All those American troops on the island of Okinawa in the fall of 1945, who were ready to invade Japan—they knew that the Japanese were ready to

[3] Ibid., 426.

die rather then to give in; I have had so many...come up to me and say that those two bombs saved their lives. I am proud that I participated in the Manhattan Project, and if anything, I often wonder how many more lives could have been saved had we done the bomb a year or so sooner.

This question has a painful personal relevance to me. My younger brother, Joe, was killed in action in October of 1944, in Italy. Shortly before he was killed, I received a letter from him in which he wrote: 'hurry up!' Although only vaguely, he knew that I was involved with war work. By the time my brother got killed, the uranium-separation plant in Tennessee was already operational but only producing grams of uranium, and we needed kilograms for the bomb. Similarly, we had produced only grams of plutonium,...again, instead of the kilograms that we needed. The idea of how to design the uranium bomb (a so-called gun-type bomb) was already more or less clear among the scientists working in Los Alamos at that time, but the idea of how to build the so-called implosion-type bomb was only starting to surface. This was eventually used for the plutonium bomb dropped over Nagasaki. I have reflected many times on my own roles in the events of those days. Could I have sped up the process? Even with a rude calculation, one can judge that had the war ended a year earlier, in mid-1944 instead of the mid-1945, possibly 15 million lives would have been saved. A heavy thought...

Getting the H-bomb took special initiative on my part. I was not simply a small cog in a big machine. We got the H-bomb only eight months before the Soviets did. If it had been the other way around, it would have been too bad for the world.

Although I am proud of my participation in the H-bomb project, it was not an easy decision to make. After having spent years on the Manhattan Project, I was happy being back in academia again. I was full of ideas. I received a grant from the Guggenheim Foundation to go to Paris to work on topics that interested me at that time. Earlier I had worked with Bohr on the liquid-droplet model of the atomic nuclei. Later, the independent-particle model of the nuclei also appeared to be a possibility, and I wanted to try to develop a unified way to describe the behavior of the nuclei. In the summer of 1949, my wife Janet and I went to Paris and I started to work there, with frequent visits to Bohr's laboratory in Copenhagen. In September of that year, we heard that the Soviets had succeeded in making their atomic bomb, and we suspected that they might have already gotten a head start in the development of their hydrogen bomb. Soon thereafter, I received a telephone call from Washington, asking me to join the H-bomb project. This was an awful dilemma for me. I thought, together with Edward Teller, that this was an obvious threat to our country. At the same time, I had just recently started to work in Paris and was enjoying it. I was struggling between my obvious patriotic duty and my love for physics. Janet and I had long discussions about it. Later, I went to visit Bohr again, and told him about all this. He asked, 'Do you think that Europe would be free of Soviet control today had it not been for the atomic bomb of the West?' Finally,...in January 1950, I decided to join the H-bomb project."[4]

---

[4] Ibid., 426–28.

When we asked him about his heroes, among quite a few others, he mentioned two ancient Greek philosophers.

"I have a stone on the island in Maine that my son and his wife brought back from the outskirts of Athens; it is from a garden where Plato and Aristotle walked and talked. And I hope some day to find a machine that…one can put the rock into, and the conversation will come out. They were certainly heroes."[5]

---

[5] Ibid., 430.

FRANK WILCZEK

I never wanted to do anything else [except science].

Frank Wilczek and Magdolna Hargittai in 2005 in Lindau, Germany. (photo by I. Hargittai)

Frank Wilczek[1] (1951–) was born in New York City. He is Herman Feshbach Professor of Physics at the Massachusetts Institute of Technology. He was awarded the 2004 Nobel Prize in Physics jointly with David J. Gross of the University of California, Santa Barbara, and H. David Politzer of the California Institute of Technology "for the discovery of asymptotic freedom in the theory of the strong interaction."[2] Wilczek earned his bachelor's degree at the University of Chicago and his master's and PhD degrees at Princeton University. His first professorial appointment was also at Princeton, followed

---

[1] I. Hargittai and M. Hargittai, *Candid Science VI*, 856–69.
[2] "The Nobel Prize in Physics 2004," *Nobelprize.org*. Nobel Media AB 2013, http://www.nobelprize.org/nobel_prizes/physics/laureates/2004/.

by one at the University of California, Santa Barbara. He joined the MIT faculty in 2000. We recorded our conversation in 2005, in Lindau, Germany.

Toward the end of our conversation, we asked Wilczek about his mentors.

"My father had an enormous influence on me, by example. He was educating himself—he was learning calculus at the same time as I was learning calculus. The fact that he was studying as a grown man was very eloquent testimony to the value of education. My mother was also very supportive. I had terrific parents. I had some excellent teachers in high school. There was then a person who had enormous influence on me in college, although I didn't realize it at the time; it came to fruition a couple of years later. His name is Peter Freund at the University of Chicago. He is a Romanian physicist and has been at Chicago for a long time. He is a very big fan of symmetry and very enthusiastic. He was also interested in things that at that time were regarded as very speculative, whereas by now they've become mainstream. He taught a course on group theory in physics, and I majored in mathematics as an undergraduate. I took his course because I was somewhat interested in physics. He taught this course, where he showed $SU(3)$ and $SU(6)$ and how they were used in physics. And suddenly I saw that they weren't just abstractions; that they had not just aesthetic value, but through them you could actually make contact with the physical world. And it wasn't ossified at all; it wasn't all 'done.' That really planted the seeds, and although I went to graduate school at Princeton for mathematics, I wasn't at all sure of what I wanted to do and I kept up with physics, and that's what led to my career. I switched from mathematics to physics when I was two years in. It was in the early 1970s, and very exciting things were going on in physics. That was... when I started to work with David Gross. I had sat in a course by David and then I joined him.

Was there a division of labor in your prize-winning work?

There was a dynamic to it. Now, it's a touchy thing and I don't want to cause any tension and touch on a delicate situation. We worked very intensely together. There were some parts to which I contributed more, and David contributed more to other parts.

Did he initiate the project?

No, I wouldn't say that. Ken Wilson had given a series of lectures at Princeton about the renormalization group that impressed a lot of people, and I sat on those and so did David. Wilson's lectures later became a book. Then there were dramatic developments in what now is called the Standard Model of electroweak interactions. David was coming more from the point of view of trying to understand the strong interactions. I was coming from the point of view of gauge theory and the weak interactions. I was interested in studying the high-energy behavior of weak interactions. The new theories avoided the famous problem of the Landau ghost that electrodynamics had, which is the question of whether there is asymptotic freedom in a different language. I brought some expertise in gauge theories to it, and David brought much more maturity and expertise in the strong interactions. He used renormalization to make concrete predictions. We both brought things together. I was a graduate student looking for a problem and discussed many possibilities. We agreed that this was terrific thing to work on.

Is the Nobel Prize changing your life?

It's definitely changing my life. My life has been very different for the last few months in terms of busyness, but it's extremely gratifying. People started treating me differently. But this is not sustainable, I can't be traveling this much. It has been very disruptive, and my ongoing projects have been put at the back burner. I'm eager to take them up again. I'm very excited about the new developments in fractional statistics, but I had to drop them because I haven't had the time to deal with them.

Will you get back to your routine?

In two weeks. I'm trying to reach a new equilibrium. It won't be the same, but I do definitely want to return to full-time research or close-to-full-time research. But I realize that it's a duty of the Nobel laureates to reach out to the public, so I am probably going to write up some of my lectures as books. I will probably work on more ambitious and riskier projects than in the past. I can afford it. I'm no longer worried about producing results that would ensure my getting tenure.

There are people who feel they have to worry about their reputation more when they are Nobel laureates.

I've thought about that, and I think it's a very unhealthy attitude. I've seen people destroyed by that. I'm not going to name names; but people who've taken that attitude have not prospered. I had time to think about this. I said to myself years ago that as soon as I get the Nobel Prize, I'm going to write some not-so-important papers just to make sure that I don't get intimidated. I'm not going to write deliberately bad papers, but I'm not going to set some unrealistic standards. That would be the way to sterility. I feel very strongly about that." [3]

---

[3] *Candid Science VI*, 867–69.

KENNETH G. WILSON

I don't think anybody expected the World Wide Web and its aftermath to happen so fast.

Kenneth G. Wilson in 2000 at Ohio State University in Columbus. (photo by I. Hargittai)

Theoretical physicist Kenneth G. Wilson[1] (1936–2013) had a background that was exceptionally fortunate for a career in science. His father was the distinguished Harvard chemistry professor E. Bright Wilson, the author of *An Introduction to Scientific Research*. His father involved Wilson in producing this book, which has served since as an inspiration and a subject of continuous inquiry for his son. Even at the age of fifty, Kenneth described himself as his late father's 'fifty-year-old graduate student.' The question was always on his mind, 'Did I meet his standards?'[2]

Kenneth Wilson obtained his PhD degree in 1961 at the California Institute of Technology, where he studied under Murray Gell-Mann. Between 1963 and 1988, Wilson was on the faculty of the physics department at Cornell

---

[1] M. Hargittai and I. Hargittai, *Candid Science IV*, 524–45.
[2] Ibid., 541.

Hans Bethe congratulating Wilson on his Nobel Prize, October 1982, at Cornell University. (courtesy of K. G. Wilson)

University. Then his interest shifted from theoretical physics to education—to the way physics is taught in schools—and he moved to the Ohio State University.

In 1980, Wilson received the prestigious Wolf Prize together with Michael Fisher and Leo Kadanoff. In 1982, Wilson was awarded an unshared Nobel Prize in Physics "for his theory for critical phenomena in connection with phase transitions."[3]

Our interaction consisted of two parts; one was a conversation in 2000 in Columbus in his university office, and the second was by correspondence. It is interesting to note that Wilson's Nobel Prize is one that might have also been shared, but it was not. In similar situations, it has often happened that the Nobel laureate took upon himself (so far, only male laureates) to justify the Nobel Committee's decision. Wilson proved to be an exception. He told us that he thought it would have been proper to share the prize, something he had never before said publicly: "Now that I'm older and wiser I realize I should've been more explicit about that."[4]

---

[3] "The Nobel Prize in Physics 1982," *Nobelprize.org*. Nobel Media AB 2013, http://www.nobelprize.org/nobel_prizes/physics/laureates/1982/.
[4] *Candid Science IV*, 530.

Wilson had interesting views about greatness in science. First in connection with the question about the greatest scientists in physics, he stressed Johannes Kepler's merits.

"In their generation Feynman and Gell-Mann were exceptional. Of more recent times, I'm not giving you names because they are still building their working careers.

It's also interesting to look at the social science situation. It is much harder to accomplish things in the social sciences because they are in the pre-paradigm phase. That's why one cannot get the same kind of recognition that one can get in a science in a post-paradigm phase or if one is the person who creates a paradigm. That's why Kepler is sometimes not considered to be in the same class as Newton even though people who have studied that period say that Kepler was even more extraordinary. I can't make up my mind about that switch, to consider Kepler more extraordinary than Newton. But he is certainly Newton's equal.... Most people who rank Newton versus Kepler have not studied the work of either. Newton published the *Principia* in 1680, and there followed three hundred years of development in terms of how one actually applies Newton's laws. One can clearly separate all the later developments from what he did himself, which was to write down the three laws, along with a small initial set of applications. In most other cases, the people who are the pioneers get an area of science started, but they don't develop anything nearly as powerful as Newton's laws have been. In these other cases, the follow-up work is of greater importance than even the follow-up work on Newton's laws has been."[5]

When we asked Wilson whether James Watson and Francis Crick could be counted among the greatest scientists, along with Copernicus, Newton, and Einstein, he laughed. So we extended the question to include his views about the importance of the double helix. Wilson responded with his own question: "You're a chemist; I'm a physicist. Do the biologists laugh? You should try that. I'll answer your question after you've tried it on some biologists." After one of us (IH) did, in 2002, we went back to Wilson. What follows is his combined response from 2000 and 2002,

"Obviously, the absolutely central role of DNA and its double helix structure in genetics is beyond question. I laugh in part because you omitted Darwin and the founders of quantum mechanics (such as Schrödinger, Heisenberg, Bohr, and Dirac) from your list. I would place all of them ahead of Watson and Crick, in terms of both the originality and creativity of their work and the breadth of their impact on our understanding of the world that we live in. But I laugh for a more profound reason. There were not all that many scientists working at the time of Copernicus and Newton, and even in the time of Darwin. But in the twentieth century, the dominant story in science is not the contributions of individuals, not even of Einstein, let alone Watson and Crick. Instead, the big story (at least for me) is the growth of institutions that are dependent on the research of increasingly large numbers of scientists, such as the agricultural experiment stations and the agricultural extension system in the US, that have transformed modern society. (A history of these agricultural institutions is provided

---

[5] Ibid., 543–44.

in Wallace E. Huffman and Robert E. Evenson's *Science for Agriculture: a Long-Term Perspective*.) The areas of the economy that have been transformed include medicine, transportation, telecommunications and information storage, energy, industrial materials, defense, large-scale construction (including skyscrapers and bridges), and media and entertainment.

All of these transformations have been accomplished with the crucial help of a growing base of knowledge: scientific knowledge, engineering knowledge, and professional knowledge in areas such as medicine and law. In fact, knowledge in toto has become the dominant source for economic wealth, social progress (if any), and military might. In my view, the person that has told this story, which is much bigger than DNA, better than anyone else to date is Peter Drucker, through a long list of books, of which *The Age of Discontinuity* (1968), *Innovation and Entrepreneurship* (1985), and *Post-Capitalist Society* (1993) are among the most profound. But Drucker is only a transitional figure. He is exceptional, yet not nearly enough so to rank with Newton or Einstein. Despite Drucker's work, and that of many other social scientists, too, the growth of knowledge from its beginnings long ago to its present dominant role in society is still much farther from being understood than are many of the well-established topics of research in the natural sciences.

Nevertheless, as I learn more about the growth of knowledge as a whole, I become less inclined to take any list of 'big names' in science seriously. I admire the accomplishments of Newton and Einstein, and of Darwin and Watson and Crick too. But I give equal weight to seemingly far more mundane matters such as the growth of reference materials (encyclopedias, dictionaries, and the like) in libraries, without which hardly anyone would have the possibility of learning about the accomplishments of the big and not-so-big names in science. My father helped me to understand the importance of reference material for scientists, as part of a whole chapter of *An Introduction to Scientific Research* on searching the literature. His comment about encyclopedias is that they 'are surprisingly useful for acquiring a first view of a new field.' I could not agree more, based on my own experience with them. More generally, I find that *An Introduction to Scientific Research* provides a far more realistic and balanced view of how science makes progress than one can learn from a list of, or a more detailed study of, the 'big names' in science."[6]

---

[6] Ibid., 544–45.

SECTION 2
Chemists

HERBERT C. BROWN

Fortunately, it is the research director who is in a position to insist upon high standards; he does not have to do the work himself.

Herbert C. Brown in 1997 in front of the plaque of the Herbert C. Brown Laboratory of Chemistry at Purdue University. (photo by I. Hargittai)

Herbert C. Brown[1] (1912–2004) was born in London, UK, and when he was two years old his family emigrated to the United States. In 1938, he received his PhD degree at the University of Chicago. In 1943, he got an appointment as Assistant Professor at Wayne State University. In 1947, he moved to Purdue University and stayed there for the rest of his life. He was both an inorganic and an organic chemist. Brown shared the 1979 Nobel Prize in Chemistry with Georg Wittig

---

[1] Istvan Hargittai (Ed. Magdolna Hargittai), *Candid Science I: Conversations with Famous Chemists* (London: Imperial College Press, 2000), 250–69.

Brown and Balazs Hargittai in 1995 in West Lafayette, Indiana. (photo by I. Hargittai)

for the "development of the use of boron- and phosphorus-containing compounds, respectively, into important reagents in organic synthesis."[2]

In the beginning of Brown's higher education, he had to overcome adversities.

"I graduated high school in 1930, and for two years I tried to find a job, without success. This was the time of the Great Depression. Then I decided to go back to school. There was Crane Junior College, operated by the City of Chicago. It did not involve payment of any tuition, but they were very selective. Fortunately, I was accepted, and I went there to become an electrical engineer, because somebody had told me that electrical engineers made good money. At that time, that was the most important thing in my life.

The first year of the electrical engineering curriculum required the study of chemistry, and the subject fascinated me. I had a very good memory and could learn these things better than anybody else in the class. After the first year, I decided that I could write a book on chemistry. I paid my younger sister five cents an hour to type it. The book was about general chemistry, my own interpretation. It's still around somewhere in my archives.

---

[2] "The Nobel Prize in Chemistry 1979," *Nobelprize.org*. Nobel Media AB 2013, http://www.nobelprize.org/nobel_prizes/chemistry/laureates/1979/.

Also at Crane, I met a young girl, sixteen years old, who was registered as a chemical engineer. At that time, this was unheard of. There were student engineers at this junior college by the hundreds, but only one was a girl, Sarah Baylen. At first, she hated me because she had been the brightest student in the chemistry class until I joined, and she didn't like to be second.

But after one year, Crane Junior College was closed because of lack of funding. There was a Dr. Nicholas D. Cheronis, originally from Greece, at this college, and when our college closed down, he did a very nice thing. He operated a small commercial laboratory at his home, in an expanded garage, called Synthetical Chemicals. He used to make costly chemical indicators and other high-priced chemicals as a side business. He invited about ten of us students to come out and do what we wanted to do in the laboratory, to keep us off the streets. Sarah was there, and I went there....

After a year, three new junior colleges were opened in the City of Chicago. Dr. Nicholas Cheronis was appointed Head of the Physical Sciences Division at Wright Junior College. I went, and so did Sarah. In the meantime, we had become good friends. At Wright Junior College, Nicholas Cheronis taught many of the chemistry courses. I took his course on organic chemistry. He was writing a book on doing organic chemistry on a semi-micro scale. He gave us all the experiments to test to make sure they were all right. That was the first book that described laboratory experiments on a smaller scale than the usual procedures.

They gave me a free hand there, and I was allowed to experiment with equipment not yet used in classes. I was permitted to publish a new paper, the *Physical Science Monthly*. It described various experiments. I was the editor and often the reporter who did all the work. They were very good to me at this college, and they told me, when I was approaching graduation, that I should take the examinations for competitive scholarships at the University of Chicago. I signed up for this, but when I saw the examination, I thought it was hopeless. At that time, the University of Chicago had a president, Dr. Robert Maynard Hutchins, who thought the way to produce ideal college graduates was by educating them in the great books, "the hundred Great Books," with the education based primarily on philosophy, psychology, literature, etc., with not much attention paid to chemistry, physics, mathematics, and so on. But all of my education had been in these specialized areas. However, I did the best I could, and to my great amazement, I won a scholarship. In 1935, I went to the University of Chicago. Another idea of President Hutchins's was that the students should not spend longer time than they needed to get their college degree. It cost no more money to take ten courses a quarter than to take the usual three. So I took ten courses per quarter and finished in three quarters, in 1936.

I didn't apply for a teaching assistantship because I wanted to marry Sarah, and in those days, one didn't get married without a job. Professor Julius Stieglitz was now emeritus. He had been head of the chemistry department for many years. He had a very famous twin brother, Alfred Stieglitz, who married an artist, Georgia O'Keefe. Julius Stieglitz was teaching the advanced organic course. Sarah and I were in that class. He used to have a custom of speaking for about five minutes, and then he would ask the class a question, a mechanism to keep them alert, I guess. I was always volunteering with the answer. Apparently, I made an impression upon him. He called me into his office, and said, "Mr. Brown, I know that you haven't applied for a teaching assistantship, which would allow you to go on for a PhD degree. Why haven't you? I am confident that you would do very well as a PhD."[3]

---

[3] *Candid Science I*, 253–56.

ERWIN CHARGAFF

Whether it is a work of art or a work of science, everything has to be consumable.

Erwin Chargaff in 1994. (photo by I. Hargittai)

Erwin Chargaff[1] (1905–2002) was born in Czernowitz, in the former Austria-Hungary (now the Ukraine), and died in New York City. When we visited him, he was Professor of Biochemistry Emeritus at Columbia University. He received his doctorate at Vienna University. He discovered the base-equivalence of DNA and that DNA was organism-specific. He earned many distinctions and was member of the US National Academy of Sciences and other learned societies. He received the National Medal of Science and many other awards. We recorded our conversation in 1994 at his home on Manhattan's Central Park West.

(IH narrates.). I have been asked repeatedly about this snapshot, which displayed an apparently happy Erwin Chargaff, whereas he was known to be a bitter person. I had been in correspondence with him before I visited him in 1994 and visited him again in 1998 with MH. In 1994, I took a few snapshots of him alone and with his wife. He was

---

[1] I. Hargittai and M. Hargittai, *Candid Science I*, 14–37

dressed in a suit and tie, of course. I was staying with a friend, also on Central Park West, close to the Chargaffs' apartment,. When I returned home, I realized that I'd used a wrong setting on my new camera. It was predigital time, but even without having the film developed I knew something was wrong. I called Chargaff right away and told him I would like to come back to take a few more snapshots. Fortunately, I succeeded in overcoming his initial reluctance. In a few minutes, I was back in the Chargaff residence. Chargaff had changed and was now informally dressed, and apparently he was amused by my inexperience as a photographer. I caught him in a smiling mood. I could not, though, persuade Mrs. Chargaff to sit for me for a second time. This is how it happened that when I first published our conversation, this solo photo was accompanied by an out-of-focus one of Erwin and Mrs. Chargaff together. When I next visited Chargaff in 1998, he expressed his gratitude to me for the out-of-focus image, not because of its poor quality, but it turned out that it was the last photograph of the two of them together. By the time of my 1998 visit, she was no longer alive.

Chargaff's most famous discoveries are that the composition of DNA is organism-specific and that the purine and pyrimidine bases are present in equal amounts. He discovered the base equivalence, but did not ask the question of why it is so. Some think that his bitterness stemmed from his realization that he had missed discovering the double helix, compounded by the fact that Chargaff despised the discoverers of the double helix, James D. Watson and Francis Crick, and said that they "made this terrific noise; this was an advertising company."[2]

However, the discovery of the base equivalence was far from trivial. The numbers characterizing the base concentrations scattered quite a bit, and it took a sharp eye to notice the pattern in those numbers and courage to come out with it. By the time Chargaff published his observation, he had already built up his reputation, and this could easily have ruined it.

Chargaff first visited the United States at the age of twenty-three. He won a scholarship to study for two years at Yale University. When the two years expired, he was offered an assistant professorship from Duke University, but the stipulation was that he would have to work on tobacco. Chargaff declined to spend his life on tobacco research. But he was a life-long smoker and in 1994 he joked, "if I'd run out of money, I could hire myself out as an advertisement for the tobacco industry."[3]

Following his return to Vienna, he went to Germany and worked in Berlin. After the Nazis came to power he left for Paris, and after a stint at the Intitut Pasteur in 1934, he returned to the United States, this time for good. He found that the allocation of research funding in pre-Nazi Germany had been very simple and efficient compared with how young scientists were competing for money in the United States.

"If I wanted to supplement what the department paid me, I should submit an application to the *Deutsche Notgemeinschaft* for a grant [it was the Emergency Association of German Science, which was the German funding agency for scientific research that

---

[2] Ibid., 25.
[3] Ibid., 18.

time]. A certain amount would then be set aside to supplement my salary, and I could buy chemicals, etc. I made an application and got an invitation in writing to see— and this was incredible—the chief of the *Notgemeinschaft*, who wasn't even a scientist. His name was Schmidt-Ott and he was an orientalist. I was let into the office of His Excellency; and he made me sit down, and we started talking about what books I was reading at that time. Then he asked me about my plans and I told him that I had written a proposal about my work on polysaccharides of tubercle bacilli, and so on. He asked me what a polysaccharide was. I explained, and then he said that I would hear from him soon. Three days later, I had the grant. This is not as stupid as it may seem. If you get the right people, they don't have to spend too much time on it. I find it silly when they call this proposal valuable and the other proposal not valuable. Most proposals are half so and half so because one doesn't know yet; and if it is good, it still may not work. So the peer reviews are completely useless, except that they grow into old boys networks. I think the most important thing is to get the general behavior, the general way of thinking of the person, rather than to decide 'this is a marvelous problem.' They are not marvelous except in lucky hands, in very gifted hands. The hands you can't look at in a proposal."[4]

---

[4] Ibid., 24.

MILDRED COHN

[Women scientists] should be scientists first and a woman scientist second.

Mildred Cohn in 2002 in Philadelphia. (photo by M. Hargittai)

Mildred Cohn[1] (1913–2009) was born in New York City. She studied at Hunter College and earned her PhD degree from Columbia University. Following stints at Cornell Medical School and the Washington University School of Medicine, she spent most of her career at the University of Philadelphia. Cohn was a biochemist and did pioneering research on the use of the isotope oxygen-18. She was member of the United States National Academy of Sciences, and a recipient of the National Medal of Science. We recorded our conversation in 2002 in her home in Philadelphia.

During the early part of her career, Cohn worked with six Nobel laureates; the first among them was Harold Urey.

"I had decided to work with him before he got the Nobel Prize and joined him a few months after his award. But I had decided that I wanted to do my graduate work with

---

[1] Istvan Hargittai (Ed. Magdolna Hargittai), *Candid Science III: .More Conversations with Famous Chemists* (London: Imperial College Press, 2003), 250–67.

Cohn with her husband, Henry Primakoff, and their three children, Laura (b. 1949), Paul (b. 1944), and Nina (b. 1942). (courtesy of Mildred Cohn)

him when I had attended his lectures—he was so inspiring. The Nobel Prize did not make any difference. He was very enthusiastic about his subject. He had profound insight in chemistry, and physical chemistry in particular. Actually, he did not like the term 'physical chemistry'; he preferred the term 'chemical physics' for his type of research. He was the first editor of the *Journal of Chemical Physics*. He started a new seminar series in chemical physics; he didn't like the content of the physical chemistry seminars. I remember when I started working on the acetone–water ^{18}O exchange. I first investigated it in the gas phase, which is what he wanted me to do. Then I thought, why not try it in solution. I did and I found that the exchange was easily measurable at room temperature. In the gas phase I had to go to 80°C to get an exchange. I went to Urey and I told him that I ought to study this reaction in the liquid phase. He said to me, 'I know something about the gas phase and I know something about the solid phase, but the liquid phase is a complete mystery to me, and I do not do experiments where I have no theory to guide me.' I told him that I thought it would be much more interesting to study the liquid phase because we could study acid and base catalysis in solution. So Urey suggested to me that I talk with Professor Hammett, who knew all about the liquid state. I did it in solution and it resulted in my thesis.

Urey was a remarkable man. Of all the scientists I ever worked with, he is my favorite. He was very naïve about social questions. It was at a time when Hitler was on the rise; I was in Urey's laboratory from 1934 to 1937. He was great friends with Professor Rabi in the physics department, and he had suggested to Rabi that he move to New Jersey, where Urey lived. But Rabi preferred to stay in New York with his own people. Urey told me that he understood Rabi's position, since Rabi was a Jew; whereas he, Urey, could probably accommodate to the fascists if they ever took over. But Urey by then had already signed a petition that Columbia University should not participate in the celebration at Heidelberg, Germany, to which Columbia had agreed to send a representative. When Urey signed the petition, he was immediately labeled as an anti-fascist. Although Urey was very naïve, he learned fast, and later he became very active in political causes. He was the only one on the faculty there who wore a Roosevelt button; all the others were Republicans. He was also very generous. He was the only one who cared about the underdogs, the graduate students. He was worried about whether they had enough support, and in particular, he worried about me. One day he said to me, 'Miss Cohn, what are you doing for money?' He had gotten me financial aid. At that time, they had something called NYA, National Youth Act, like a work study program for students today. He said the only thing he wanted me to work on was my own thesis. The aid was available only during the academic year. In the summer, he said to me, 'Miss Cohn, why don't you let me lend you some money? Ever since I got the Nobel Prize, I have wanted to use some of it to help my students. Some day, when you have a job, you can pay me back.' There aren't many professors who would do that, and certainly not today. He never forgot that he had been a poor boy himself, a farm boy. He told me once that the only reason he had ever gone to high school—there was no high school where he lived, so he had to go and board somewhere—was due to his uncle who died and left him $300 That's how he managed to go to high school."[2]

[2] Ibid., 255–57.

JOHN W. CORNFORTH

I see the origin of life as a minor victory against decay.

John and Rita Cornforth in 1997 in Lewes, England. (photo by I. Hargittai)

John W. Cornforth[1] (1917–2013) was born in Sydney, Australia. He was an organic chemist interested in the biological aspects of chemistry. He entered Sydney University at the age of sixteen. He was an excellent student, despite the heavy challenge of being deaf. In 1939, Cornforth won an "1851 Exhibition" scholarship to work at Oxford University, UK. There were two such scholarships awarded every year to students chosen from the six Australian universities. The other awardee in 1939 was Rita Harradence, who was one year ahead of Cornforth in her university studies. Both went to work with Robert Robinson in Oxford. They married in 1941. Harradence became Cornforth's

---

[1] I. Hargittai and M. Hargittai, *Candid Science I*, 122–37.

wife, partner, and often helper as a bridge to the outside world. We recorded our conversation in 1997 in Lewes, England.

On his deafness.

"I was getting deafer all the time. I could not use the hearing aids that were available then because the sound came out distorted. I did not use lip-reading very much at the time. Even if I had been an expert in lip-reading, I could not have used it for lectures because lip-reading is a guessing game. You always have to interpret what you think is the content. It is not good for learning new ideas. I have to tell that to deaf children, who sometimes ask for my advice about how to get through university. They have more help now."[2]

Cornforth participated in the penicillin project.

"We arrived in Oxford in 1939. Just about that time, people were already working there on extending Fleming's work on penicillin. The trouble was that penicillin is very unstable in its crude form. Fleming had asked a chemist named Raistrick to see whether he could get anything out from the broth, but this attempt was unsuccessful. However, Chain, another chemist, knew about the sensitivities of penicillin, and he, with Florey and others, took up this work. They devised a beautiful assay to measure the quantity of penicillin in arbitrary units. They were then able to find out what conditions wouldn't inactivate the stuff and then, gradually, to concentrate it. Fleming's original culture was not a good producer by modern standards. So production had to be on a large scale, because large volumes were needed to scrape together enough penicillin for a treatment. The first time they challenged the staphylococcus in vivo, it was a spectacular success. The infection killed all the controls in a few hours, and all the treated animals survived. That was in 1940.
When was the first human case?
It was an Oxford policeman, in 1941. He had developed septicemia, blood poisoning, from an untreated sore. They did not have much penicillin. If they had given him all that they had in one dose, they might well have cured him. However, they decided to divide it up. Upon the first injection, his temperature went down and his blood cleared. But then there was a relapse. They gave him another injection and the same thing happened. This went on until none was left, and the man died.
Rita and I entered the penicillin project in 1943, when the structure of the molecule was being determined. We were able to synthesize D-penicillamine, a fragment representing nearly half the molecule. The chemistry occupied hundreds of chemists here and in the US during the rest of the war. There was a monograph published on the combined efforts, *The Chemistry of Penicillin*, Princeton University Press, 1949. I wrote one of the chapters."[3]

---

[2] Ibid., 125.
[3] Ibid., 126.

Cornforth wanted to share the lesson of his experience.

"If you have the predicament of deafness, you cannot get anything much from meetings or from conversations involving more than two people. This is why I tend to go for the printed record. I disbelieve in abstracts and reviews automatically. An abstract is what the abstractor thought was important, or what an editor and his referees allowed to appear of what the author thought was important! By the time you read either a review or an abstract, it has been filtered through too many brains. The best you can do is to go for the original paper, but what I get from a paper is almost never what was intended. So I am worried about the tendency to avoid full publication of results and to rely on databases, while access to the whole of the literature is becoming more difficult because libraries cannot afford full coverage. The quantity of information being produced is stupendous, but the quality is nobody's business. The usual 'preliminary' publication is almost all interpretation, not the actual experiments, which are the only things of lasting value. The problem can be solved only on a multinational scale, and I should like to see a European central library whose readers can call up any paper, however old or obscure the source, on their computer screen and decide if they wish to copy it. For that reason among others, I take a lot of care with my papers, and I am content that they will be my memorial. 'What you did, why you did it, and what were your results' is as valid an ideal as when Rutherford enunciated it."[4]

---

[4] Ibid., 133–34.

DONALD J. CRAM

This year's gamble in research is next year's conventional wisdom.

Donald J. Cram in 1995, Palm Desert, California. (photo by I. Hargittai)

Donald J. Cram[1] (1919–2001) was born in Chester, Vermont. He studied at Rollins College in Florida and the University of Nebraska–Lincoln, and earned his doctorate degree at Harvard University. For forty years, until his retirement, he was on the faculty of the University of California, Los Angeles. At the time of our conversation he was Professor Emeritus in the Department of Chemistry and Biochemistry of the University of California, Los Angeles (UCLA). He was a co-recipient of the 1987 Nobel Prize in Chemistry together with Jean-Marie Lehn and Charles J. Pedersen "for their development and use of molecules with structure-specific interactions with high selectivity."[2] We spoke to him in 1995, at his home in Palm Desert, California.

---

[1] I. Hargittai and M. Hargittai, *Candid Science III*, 178–97.
[2] "The Nobel Prize in Chemistry 1987," *Nobelprize.org*. Nobel Media AB 2013, http://www.nobelprize.org/nobel_prizes/chemistry/laureates/1987/.

The early loss of his father determined Cram's path to adulthood and impacted him throughout his later life.

"My father died when I was not quite four years old. He had been a successful lawyer in Canada, who emigrated with his family of my mother and three girls to the United States to manage a citrus venture he and friends had invested in in Florida. The venture failed, and the family moved to Vermont, where my father tried to become a farmer, about the time I was born. Shortly thereafter he died, leaving my mother with four girls and me. The oldest child was thirteen years old. The absence of any male in or close to the family led me from an early age to seek a father image, first from the many books I read, and later in the teachers and scientists I encountered. I was in my mid-thirties before an inner core of values was in place, and my character was formed. It was a composite of what I had admired in people, I had read about or had come to know. Possibly even of greater importance were my many encounters with people who illustrated what I did not want to resemble. In retrospect, I judge that what are ordinarily considered to be misfortunes provided me with the advantages of being able to test myself against circumstances and to grow in confidence, skill, and judgment with each encounter."[3]

Once Cram had come to the idea of building his new compounds, he went around giving talks about it as if—in his words—"raising the challenge" and his courage. Once he talked about his ideas in public, there was no way back; he had to do what he had talked about. In addition to creating a new chemistry, he was also very good in creating new names for it and for the kinds of new substances he was aiming at. Initially, it was "host-guest chemistry"—meaning that a larger host molecule would house a smaller guest molecule. Later he often used the term "container chemistry," meaning that a larger container held a smaller molecule.

"The words container chemistry we apply mainly to capsular complexes such as our spheraplexes and carceplexes, in which decomplexation is mechanically inhibited. The term 'molecular container' is more suggestive of geometry than the more general term 'host.' Good nomenclature elicits images and aids reasoning by analogy, the organic chemist's 'best friend.'

The evolution of our thinking that led to container compounds was very simple. We asked ourselves what geometric feature was common to nature's active sites, particularly the enzyme systems and RNA. The answer was concavity. The simplest way to get cooperativity between catalytic or binding sites is to imbed functional groups in an enforced concave surface, so that they converge. Multiple and simultaneous contacts between substrates and receptor sites are required to provide the binding free energies required for the molecular machinery created by evolution. The crystal structures of nature's complexes provided this information. A literature search in the 1960s revealed that only a few simple organic compounds containing enforced concave surfaces were known, examples being cholic acid and the cyclodextrins. Both were known

[3] *Candid Science III*, 179–80.

to complex organic compounds. Accordingly, we established a research program for designing, synthesizing, and studying the binding properties of organic compounds containing enforced concave surfaces shaped like saucers, bowls, and vases. We called these compounds *cavitands*, and their complexes *caviplexes*. The suffix *-and* was derived from the word 'ligand.' The family name *host* was applied to the complexing partner with the *concave* surface, and *guest* to the partner with a *convex surface*. Once cavitands were in hand, it was inevitable that they would be attached together at their rims to compose closed-surface compounds with large enough enforced interiors to imprison guest molecules. We call such hosts *carcerands*, from the Latin word *carcer*, meaning prison.[4]...

The central idea of the research was that the exquisite chemical activities of biological processes depended largely on complexation involving large numbers of weak but additive attractions and that enzymic catalysis, immune responses, and genetic information storage, retrieval, and replication might all be modeled. We would use simpler host frameworks than those composed of amino acids, sugars, phosphate esters, and the heterocycles of the natural world. The art lay in developing rigid frameworks containing binding or catalytic sites pre-organized to act cooperatively on particular guest compounds. In effect, by complexation we hoped to turn ordinarily remote, uncollected, and nonoriented functional groups into arrays of pre-organized, neighboring, and cooperating groups. This was an ideal field for me. I have a tolerance for ambiguity, for failure, for groping, coping, and gambling that this field certainly exercised. This year's gamble in research is next year's conventional wisdom."[5]

---

[4] Ibid., 189.
[5] Ibid., 193.

PAUL J. CRUTZEN

Chemical research is very important for protecting the environment and especially in the recycling business.

Paul J. Crutzen in 1997 in his office at the Max Planck Institute for Chemistry in Mainz, Germany. (photo by I. Hargittai)

Paul J. Crutzen[1] (1933–) has been associated with the Division of Atmospheric Chemistry of the Max Planck Institute for Chemistry in Mainz, Germany, since 1980. He was born in Amsterdam, earned his scientific degrees in Stockholm, and has been associated with a number of research institutions in the United States. He studied civil engineering and received his PhD degree in meteorology. In 1995, he received the Nobel Prize in Chemistry. He shared the award with F. Sherwood Rowland and Mario J. Molina "for their work in atmospheric chemistry, particularly concerning the formation and decomposition of ozone."[2] We had our conversation in 1997 in Mainz.

---

[1] I. Hargittai and M. Hargittai, *Candid Science III*, 460–65.
[2] "The Nobel Prize in Chemistry 1995," *Nobelprize.org*. Nobel Media AB 2013, http://www.nobelprize.org/nobel_prizes/chemistry/laureates/1995/.

When we asked Crutzen how much "chemistry" he knew, he laughed heartily and responded: "This is a mean question." But of course, he learned what he needed and received assistance from others when needed.

"Chemistry is an enormously broad field. Atmospheric chemistry is a part of it. I feel most at home in photochemistry, radical chemistry, spectroscopy, and quantum chemistry. Wherever I am weak, we have others as excellent specialists and I support them. We need a lot of analytical chemistry, for example. Reaction simulation is also very important for atmospheric chemistry, and so is gas-phase kinetics, and more recently, surface reactions and atmospheric organic chemistry. We study the breakdown of hydrocarbons emitted by vegetation.

A very important part of my current research is tropical chemistry. The biosphere influences the chemistry of the atmosphere in a fascinating way. Furthermore, in less than fifty years the most important impact on the atmosphere by human activities will come from the tropics and subtropics. However, for organic gases we must take into account the input by nature. The global production of hydrocarbons from trees is a factor at least ten times higher than the emission of hydrocarbons from fossil fuels. There is a high potential to create photochemical ozone smog, especially in the tropics and subtropics. Another issue is the impact of burning of waste, dry grass, bushes, and forests in the tropics during the dry season. Of course, we can't tell people there what to do, but this practice deteriorates the environment and has an enormous influence on the chemistry of the atmosphere. I have been interested in these questions since the end of the 1970s. We carry out large-scale experiments in the region, involving a lot of aircraft of several countries. We're at the stage of collecting scientific information about the relationship between natural and anthropogenic processes in the atmosphere....

Chemical research is very important for protecting the environment, and especially in the recycling business. We should be very serious about the potential consequences of what we produce, from the beginning to the end product and beyond. The CFCs [chlorofluorocarbons] were once considered to be safe, the ideal, nontoxic substances, but the consequences weren't thought through to the end. In the future, we should be more thorough. This is both to the advantage of mankind and to the advantage of the chemical industry."[3]

---

[3] *Candid Science III*, 465.

JOHANN DEISENHOFER

Yes, among friends. [Deisenhofer, answering a question about whether he talks freely in Texas about having lost his faith.]

Johann Deisenhofer (*on the left*) and Hartmut Michel in 2001 in Stockholm. (photo by I. Hargittai)

Johann Deisenhofer[1] (1943–), Hartmut Michel (1948–), and Robert Huber (1937–) shared the Nobel Prize in Chemistry in 1988 "for the determination of the three-dimensional structure of a photosynthetic reaction centre."[2] At the time of their award-winning research, Deisenhofer and Huber were in the Section for Structure Research of the Max Planck Institute for Biochemistry in Martinsried (near Munich), Germany. Michel was in a different section of the same institute.

---

[1] I. Hargittai and M. Hargittai, *Candid Science III*, 342–53.
[2] "The Nobel Prize in Chemistry 1988," *Nobelprize.org*. Nobel Media AB 2013, http://www.nobelprize.org/nobel_prizes/chemistry/laureates/1988/.

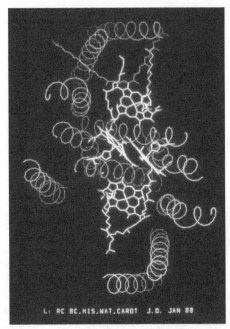

The structure of the photosynthetic reaction center with approximate twofold symmetry. (courtesy of J. Deisenhofer)

Huber has retired; Michel is at the Max Planck Institute for Biophysics. Deisenhofer is at the University of Texas Southwestern Medical Center in Dallas, Texas. We spoke with Deisenhofer in 2000 in his office in Dallas.

The most interesting part of our conversation was when Deisenhofer described the moment of discovery. At the time of their work their target—a membrane protein—was the largest biomolecular system to date whose structure had been elucidated. Such work consists of many components. Perhaps, the most difficult portion is the production of a single crystal of the membrane protein suitable for X-ray crystallographic structure determination. But the most spectacular moment is when the structure emerges. This is what Deisenhofer experienced.

"The most exciting stage in our work is when you get the electron density distribution of the molecule but there is no model yet. Then you are building an atomic model into that electron density distribution. The electron density distribution is visualized as a set of contour lines that outline a globular object, which has a very confusing internal structure. The photosynthetic reaction center is a huge molecule of ten thousand non-hydrogen atoms.

I began to look through, trying to decide where to start and what features could be assigned with some certainty. Then I learned to recognize chlorophyll molecules; I hadn't worked with chlorophyll before. There were also hemes that were better known. I found there all the features that people who had been doing, for example, spectroscopy had described. All available physical chemical methods had been applied to analyzing the composition of the reaction centers. It was extremely exciting to localize these features and build models for them. When I stepped back to see the arrangement, the unexpected observation about it was that it was symmetric. There was a symmetry in the arrangement of the chlorophyll that nobody had anticipated. Nobody, to this day, completely understands the purpose of this symmetry. I think it can be understood only on the basis of evolution. I think that the photosynthetic reaction started out as a totally symmetric molecule. Then it turned out to be preferable to disturb its symmetry, sticking to an approximate symmetry but changing subtly the two halves of the molecule. Because of the difference in properties of the two halves, the conclusion had been, before the structure came out, that there cannot be symmetry; that it has to be an asymmetric molecule. Now, when people looked at the structure, it looked totally symmetric to the naked eye. That realization was the high point I will never forget.

The technical device of model building of that time, the computer display, was much inferior to what we can have today. It allowed only a small part of the whole molecule to be looked at, at one time. So you were building details and little things, and every now and then, you summed up these details and looked at the bigger picture at a lower level of detail. Those were the most exciting moments in the whole project for me.

Were you alone when you noticed the symmetry for the first time?
I was. I was sitting in a dark room and I was interacting with the machine.
What did you do at that moment?
I started smoking again. I'd quit smoking, but I could not live with this excitement without doing something like that. It was a bad decision because it was very difficult to stop smoking for a second time. It lasted, the smoking, that is, for half a year.

I also called Hartmut and showed him the whole thing. It was very nice to have a colleague like him, who in many ways complemented my expertise. We could give each other many things in the course of this work. It was a relationship of complete trust. When such a story becomes known, there is always a temptation to claim all the credit. This did not happen between us, and I'm very glad that it didn't because it could have ruined everything.

Hartmut pioneered a lot of things during that time. For example, he was the first to begin DNA sequencing in Martinsried. He realized that if we were to have a complete model, we needed to sequence the protein. The sequence of the proteins in other species of photosynthetic bacteria was known, and it was estimated that our molecule was about 50 percent the same. It also meant that it was 50 percent different. He would come with pieces of the new sequence and we sat together and tried to find it in the electron density map. First, I'd built the model without knowing the sequence and, of course, I missed some amino acids and built some more that were not there. That was a very interesting and exciting time. In retrospect, it was the highest moment of excitement in science I ever reached; it never came again in the same intensity."[3]

---

[3] *Candid Science III*, 349–51.

CARL DJERASSI

The mission is [bridging the gulf between the two cultures].

Carl Djerassi next to his image in 1996 in his home in San Francisco. (photo by I. Hargittai)

Carl Djerassi[1] (1923–), a most versatile American chemist and lately playwright, was born in Vienna, Austria. He arrived in the United States via Bulgaria—his mother was Viennese and his father Bulgarian—and Djerassi was escaping from the Nazis. He had excellent chemistry teachers in the American colleges he attended. He received his PhD degree at the University of Wisconsin in 1945. He worked for Ciba Pharmaceutical Co. in New Jersey, and then for Syntex in Mexico City. In 1952, he entered academia, and since 1959 has been at Stanford University. We recorded our conversation in 1996 in Djerassi's home in San Francisco.

Djerassi has been most creative in science; in natural products chemistry, in the application of physical techniques, and in the synthesis of oral contraceptives. He has also been a most prolific author. We asked what he considered his most important scientific achievement.

"Importance is a subjective term.... From a societal point of view, the first synthesis of an oral contraceptive is without any doubt the most important. This was almost

---

[1] I. Hargittai and M. Hargittai, *Candid Science I*, 72–91

forty-five years ago, October 15, 1951. It's a very specific date, recorded in our lab notebook when we synthesized the steroid that was the active ingredient of oral contraceptives, which were eventually taken by hundreds of millions of people. There are very few synthetic drugs that have been around for forty-some years. I don't really know if there has been any more important discovery from a societal standpoint during the postwar period."

He had written in the 1980s that not much change could be anticipated in this respect.

"To my regret, I was 120 percent correct. Take the ten biggest pharmaceutical companies in the world: not one of them is working on new methods of birth control; none of them is even selling any birth control agent. Birth control is just not on the high priority list of the pharmaceutical industry. The pharmaceutical industry is interested in making money, and I'm not blaming them for that, but they focus their activities on the wealthy parts of the world, that is, Japan, North America, and Western Europe. This northern tier of countries are also the geriatric countries, where nearly 20 percent of the people will soon be above the age of sixty. I call the southern tier of countries the pediatric countries where about 45 percent of the people are below the age of fifteen. Their health priorities are totally different. The health problems of the geriatric societies are not birth control....

What then do the pharmaceutical companies focus on? They focus on inflammation because that is the most common disease of an aging population....Cardiovascular drugs are another one, next is cancer, Alzheimer's disease. These are very important diseases, and they are high-profit ones, and for them people are willing to accept risks. For birth control, people aren't willing to take any risk."

He earned a lot of recognition, and many share the opinion that he had produced enough results for three Nobel Prizes. We asked him how he felt about this.

"I feel exactly the same way as do most scientists. It'd be great to get it. But paraphrasing a Swedish scientist involved with the Nobel Committee at one time, the Nobel Prize is marvelous for science and terrible for scientists.

I have been present at the Nobel Prize ceremonies and was even present once at the deliberations of the Committee. I'm a foreign member of the Royal Swedish Academy of Sciences, so I have the right to nominate each year. So I've had quite some experience with them, but most of it, unfortunately, after I'd written my first novel, *Cantor's Dilemma*, although I wouldn't have changed much of the plot.

In *Cantor's Dilemma*, I write in great detail about the Nobel ceremony....[In the book] I made up a conversation the Queen would have with her neighbor....In that conversation, I had the Queen comment on...how Americans eat: they cut with knife and fork, and then they put the knife down, switch the fork, and use the fork only. (I myself eat like a Central European and find it ridiculous to waste all this time by switching the fork.) Then she goes on commenting on how people eat peas....The Americans chase the peas with their forks and they haven't got anything to help them getting on their forks, while the Central Europeans just shovel them with their knives onto their forks. But then there are the English, who hold their forks inverted, so that no peas can be kept on them. So you can divide the world between the British and ex-British colonials who eat the peas in this silly way, and then the Americans who eat them in this counterproductive way, and the Central Europeans.

A few years later, when I gave a lecture at the Swedish Academy of Sciences, they had a dinner for me, where the president of the Academy offered a formal toast, and presented me a Nobel 'Peas' Prize, which was a can of peas wrapped in the Swedish flag. This showed me, too, that people read my novel. *Cantor's Dilemma* has been translated into many languages, most recently into Chinese.

I also give a lot of lectures on the culture and behavior of scientists. I'm now returning to your question, how I feel about the Nobel Prize. I call these lectures 'Noble Science and Nobel Lust: Disclosing Tribal Secrets.' The desire is always there to win recognition and to win prizes, and the Nobel is just one of them, although it is the ultimate one. If you ask the Nobel laureates quite honestly, do you want to win another one, they say, yes! This desire is a fuel but it's also a poison. How to learn to maintain a balance and not let the poison overcome the fuel part is something that a number of scientists have not been able to solve. And this is getting worse these days with the competition getting increasingly brutal, and detracting from the elegance and even from the pleasure of scientific research."[2]

About being an author.

"If you'd interviewed me five years ago, I would have said that I was a chemistry professor who was also writing fiction. Today, I'm a novelist who is still a professor of chemistry, and that is a big difference. During the summer, we live in London, and it's then and there that I write my fiction. There, I have no contact with academia, I work every day, seven days a week, seven to eight hours a day. I am a full-time fiction writer in the summer.

Do you envision a mission for yourself in bridging the gulf between the two cultures?

This is *the* mission. This gulf is one of the most important social problems today, the gulf between the scientifically literate constituency, which is a very small portion of the population, and the intelligent literate community, which is scientifically totally illiterate. This is also part of the reason for chemophobia in contemporary society. The important factor, of course, is the readership, and this is why I decided to use fiction. I call it 'science-in-fiction' because I'd like to smuggle concepts of scientific culture of behavior into the conscience of people who are not interested in science."[3]

---

[2] Ibid., 79–81
[3] Ibid., 90.

GERTRUDE B. ELION

Our compounds were medically important but they were not making a lot of money.

Gertrude Elion in 1996 in her office in Research Triangle Park, North Carolina. (photo by I. Hargittai)

Gertrude B. Elion[1] (1918–1999) was a leading researcher of drug discovery first at Burroughs Wellcome and then at Glaxo Wellcome, in the changing world of drug companies. She earned a bachelor's degree in chemistry from Hunter College and a master's degree from New York University. Her research had resulted in the discovery of important drugs, known to the general public as Purinethol®, Thioguanine®, Zyloprim®, Imuran®, and Zovirax®. In 1988 she was awarded the Nobel Prize in Physiology or Medicine together with George Hitchings—her one-time mentor—and James W. Black. It was not for the discovery of particular drugs, but "for their discoveries of important principles for drug treatment."[2] We recorded our conversation in 1996 in Research Triangle Park, North Carolina.

---

[1] I. Hargittai and M. Hargittai, *Candid Science I*, 54–71
[2] "The Nobel Prize in Chemistry 1988," *Nobelprize.org*. Nobel Media AB 2013, http://www.nobelprize.org/nobel_prizes/chemistry/laureates/1988/.

During the development of the immunosuppressant Imuran, the first successful organ transplants were performed on dogs. This photo is from the time of that work. From left to right, the canines: Tweedledum, Tweedledee, Titus, and Lollipop; and the humans: Roy Calne (second from the left), Elion, George Hitchings, Donald Searle, E. B. Hager, and Joseph Murray. (courtesy of Gertrude B. Elion and Katherine T. Bendo Hitchings).

Elion noted the changing atmosphere in drug research when we asked whether she felt researchers today have the same latitude she had at the time.

"Probably not, and especially not in large companies, which are usually divided up into departments. It may still be possible for small companies. However, we always had to do everything ourselves in those early days. Although there were some pharmacologists there, they were involved with antihistamines and muscle relaxants. I could not go to them and say, 'Could you help me find out what happens to 6-mercaptopurine in the body?' So I decided to make radioactive 6-mercaptopurine, put it in the mouse, collect the urine, and find out what happened to the drug. I synthesized the metabolites, tried to find out how the compounds worked, and analyzed whether they were converted to nucleotides. It took me years to follow the metabolism in animals and then in man because the technology to separate and purify the metabolites just was not there."

Did the company support you all the time?

"Oh, yes. This was because the company was very different. Burroughs Wellcome was owned by a trust. Originally, Silas Burroughs and Henry Wellcome were two American pharmacists who went to England to make their fortune. They remained in England and formed a company in 1880. Burroughs died fairly early, and, because it was a partnership, everything was left to Wellcome. Wellcome became very interested in tropical medicine. He also became very wealthy. Since his wife had left him, he did not leave her anything in his will. He left everything to a trust, which was to have five trustees, including a scientist, a lawyer, and a businessman. They had to decide how to spend the money that the company made. There was nobody looking to make a personal profit but only for ways to give the money away in philanthropy. Thus, there was nobody looking over our shoulders.

The company, when I joined it, made most of its money from a drug for headaches. Then one of the chemists in England made an antihistamine. They made a lot of over-the-counter drugs; they made ointments; they isolated digoxin from digitalis leaf; they got into antibiotics; they were very heavily into vaccines. They were the suppliers of vaccines for the British Army during the war. They also made polio vaccine after the war. These were their main products; they did not have a lot of synthetic organic chemistry then.

Our compounds were medically important but they were not making a lot of money. You could not make a lot of money on 6-mercaptopurine; it was very useful for treating childhood leukemia, but that was not a major product. It was not really till we got into antibacterials, for example, trimethoprim, which was one of the diaminopyrimidines, that the company began to see some profit.... Then, with allopurinol, which was a direct result of trying to prevent 6-mercaptopurine from being oxidized, gout became a major market."[3]

On the difficulties women have in reconciling work and family.

"I know that if you are a woman and you want to have both a family and a career, it is very difficult. I did not have that problem, I did not have a family, although it is not the way I had planned it. My fiancé died in 1940 of a disease that could have been cured by penicillin, which was discovered two years after he died. I think that encouraged me to stay in research. It was a feeling that there were still so many things to do, that research is worthwhile and that here you can really make a difference.

You can imagine how I feel when somebody comes to me after I have given a lecture somewhere and says, 'I have a kidney transplant thanks to you and I have had it for twenty-five years.' I get many thank you letters and I keep every one of them. They represent the real reward, making sick people well."[4]

---

[3] *Candid Science I*, 62–64.
[4] Ibid., 70.

ALBERT ESCHENMOSER

The emergence of combinatorial chemistry was the transplantation of the evolutionary mechanism into chemistry.

Albert Eschenmoser in 1999 in Zurich. (photo by I. Hargittai)

Albert Eschenmoser[1] (1925–) was born in Erstfeld, Uri, Switzerland. He is Professor Emeritus of the Swiss Federal Institute of Technology, ETH, in Zurich. He studied and has spent his career at the ETH. He is kown for some tour de force chemical syntheses, such as Vitamin B_{12}. He is a foreign member of the US National Academy of Sciences and the Royal Society (London). Among his distinctions, he has received the Robert A. Welch Award in Chemistry (1974), the Davy Medal of the Royal Society (1978), the Arthur C. Cope Award (1984), and the Wolf Prize of Israel (1986). We recorded our conversation in 1999 in Zurich.

---

[1] I. Hargittai and M. Hargittai, *Candid Science III*, 96–107.

The science of DNA emerged as one of the determining research directions in the twentieth century, yet it seemed that the chemistry community was not quick enough to recognize its importance.

"During the last ten years or so, Vlado Prelog and I very often had lunch together. In discussing with him God and the world and, of course, matters of chemistry, such as the past and the present of natural products chemistry, I more than once gently provoked him by saying: 'Vlado, every year during which we did *not* work on DNA was a wasted year.' This, of course, is an overly drastic statement and, obviously, an exaggeration. However, there is a kernel of truth in it. This becomes evident when you look at DNA today and consider the impact of this single type of natural product structure on essentially everything that concerns human existence, an impact incomparable to that of any of the organic chemist's traditional natural product families, vitamin B_{12} obviously included.

At its origin, organic chemistry was meant to be the branch of chemistry that had to deal with carbon-containing compounds produced by living matter. After the explosive proliferation of the field toward the end of the last century, studying the chemistry of naturally occurring carbon compounds became the task of organic natural products chemistry. To the natural products chemists of today, it is, in a way, sobering to realize that their science in the past went along studying all those wonderful alkaloids, carbohydrates, natural dyes, terpenes, vitamins, and hormones, yet did so by virtually ignoring proteins and nucleic acids. Sobering it is, in retrospect, to be aware of that immense disparity in relevance between this single type of natural product structure, DNA, and all those natural products that have been at the center of the organic chemist's attention for more than 100 years. Furthermore, it is sobering to realize how late in the game organic natural products chemists started worrying about the chemistry of DNA.[2]...

In those lunches with Prelog, we often spoke about the early 1950s. While discussing that relevance gap between the organic chemist's natural products and DNA, I also asked him: 'You and Ruzicka, leading and powerful natural products chemists as you were at that time, why did you ignore DNA?' This was not, of course, meant to be a criticism of him, no. Yet the question seemed to me intriguing: why did organic natural products chemists ignore the DNA problem? After all, Avery had published the paper about DNA being the transforming principle, the genetic material, back in 1944. Prelog said that it was out of the question; one couldn't possibly do proper chemistry with it. One didn't know whether it's a large molecule, one couldn't isolate it, one couldn't purify it to the standards of organic chemistry; it was not a proper object for chemical research. I repeatedly asked him to write this down, because of his being a witness to that time, a witness to that attitude. I was gently prodding him, but he told me repeatedly, 'I don't like to do it because it could be misinterpreted.' However, I was increasingly intrigued by the question and thought that it was important to have his words in written form as the historical record of a witness. For about a year, I didn't get anything from him but then, one day, he came into my office and gave me a piece of paper which contained a (shortened) reply, written with a distinct touch of irony."[3]

[2] Ibid., 102–3.
[3] Ibid., 103–4.

Here is the English translation of Vladimir Prelog's statement.

Zurich, October 3, 1995

Dear Albert,

For some time you have prodded me to tell you why the great Leopold and I did not recognize, in a timely fashion, that the nucleic acids are the most important natural products, and why we wasted our time on such worthless substances as the polyterpenes, steroids, alkaloids, etc.

My light-hearted answer was that we considered the nucleic acids as dirty mixtures that we could not and should not investigate with our techniques. Further developments were, at least in part, to validate us.

As a matter of fact, for personal and pragmatic reasons, we never considered working on nucleic acids.

Yours,
Vlado

On the impact of combinatorial chemistry.

"Combinatorial chemistry is far more than just a new methodology. It is changing the chemist's thinking in a more fundamental way, it widens his outlook on nature, it brings him closer to one of the central concepts of natural science as a whole, the principle of Darwinian evolution. The emergence of combinatorial chemistry was the transplantation of the evolutionary mechanism into chemistry. Life, as we know it, is a chemical life, and one of the prerequisites of both its emergence and its evolution was the chemical world's immense diversity with regard to its molecular structures as well as their chemical and physical properties. Biological evolution proceeds by 'happening and selecting itself by surviving,' combinatorially and not by design, Darwinian and not Lamarckian.

Interestingly enough, deep down in the minds of synthetic chemists—and I conjecture that this may apply to scientists in general—is the belief that the hidden final goal of everything we pursue in science is to eventually be capable of doing things by design. Design is the ideal at which we seem to be driven to aim. If chemists today accept combinatorial methods because they have to admit that such methods offer a faster way to progress in solving certain problems, in the back of their minds they may still feel that, in the long run, the purpose of combinatorial chemistry is to speed up the progress toward achieving that ultimate goal of being able to solve problems by design."[4]

---

[4] Ibid., 104–5.

KENICHI FUKUI

The distances among human, social, and natural sciences are becoming shorter and shorter.

Kenichi Fukui in 1994 in Kyoto. (photo by I. Hargittai)

Kenichi Fukui[1] (1918–1998) was born in Nara, Japan. Fukui studied in the Department of Industrial Chemistry at the Kyoto Imperial University, from which he graduated in 1941. His first job was at a fuel factory of the Japanese Imperial Army. He began his academic career in 1943 in the Fuel Chemistry Department of Kyoto Imperial University, rising to the rank of professor in 1951, and retiring in 1982. In 1981, Fukui and Roald Hoffmann shared the Nobel Prize in Chemistry "for their theories, developed independently, concerning the course of chemical reactions."[2] Many other awards followed, and in 1988, Fukui was named director of the newly created Institute for Fundamental Chemistry.

Our conversation with Fukui took place in 1994 at his institute. It was an animated exchange. Fukui appeared broadly informed, demonstrated a good

---

[1] I. Hargittai and M. Hargittai, *Candid Science I*, 210–21.
[2] "The Nobel Prize in Chemistry 1981," *Nobelprize.org*. Nobel Media AB 2013, http://www.nobelprize.org/nobel_prizes/chemistry/laureates/1981/.

sense of humor, and spoke impeccable English to the extent we could judge it, but nothing was recorded on this occasion. Our questions had to be mailed to him prior to the meeting and his prepared responses were handed over to us in writing at the meeting. Additional questions arose in our conversation during the meeting, and the written responses to these questions were delivered to us while we were still in Kyoto a few days later.

We asked him what his suggestions for determining excellence in research were..

"The evaluation of achievements is always a difficult task in science policy. Truly original works are frequently disregarded by the majority of scientists. Obviously, however, hardly understandable works are not always good ones. Citation is no doubt an index that may be used to judge the value of research, but it is apt to be dominated by fashion and current, so that the number of citations is strongly dependent upon the field. The judgment of specialists may generally be of significance, while in some cases the intuition or farsightedness of talented non-specialists may be more reliable than the judgment of the mediocre majority in the field.

How much disadvantage is there for non-English-speaking scientists to get their papers published in the best journals and to get recognition for their work?

Nowadays the circulation of a journal is one of decisive concerns for the contributors. Non-English journals, even those written in other European languages, are not favored by the international community of scientists. Therefore, the disadvantage for non-English-speaking scientists like myself is obvious. Additional efforts are usually required to get public recognition for their work. Evidently, this is a big handicap! It should be got over through international intercourses of all sorts.

How do you view the separation of the sciences and the humanities in our times? How can we decrease the gap?

The distances among human, social, and natural sciences are becoming shorter and shorter. This comes from both progress and necessity. The former reason is evident. The progress of physiology and medicine correlates with new methods for psychology and the science of the human mind. The mathematical and computational progress in statistics greatly influences economics, sociology, and politics. Physicochemical sciences promote the progress of archaeology. The examples are numerous.

The latter reason is more important. It comes from the self-accelerating character of science and technology recently becoming appreciable. Science produces new technology, while the results of new technology promote the progress of science. The human native desire intervenes between them. The mutual acceleration of science and technology in this way brought about conveniences and comfort to human life. On the other hand, the uncontrolled development produced serious problems. Nature on the earth and environment of the earth were seriously changed, and terrible inequality in the distribution of civilization on the earth was caused. To affect these circumstances by controlling the perpetual desire of human beings for growths and advance, the cooperation of natural, social, and human sciences is a necessary prerequisite.

How should we be getting prepared for the next century in education?

The generations living in the next century must carry burden to ameliorate the conditions in which the earth is presently put. For this purpose, they have to make the level of science much higher and apply it to amend the situation, and simultaneously they have to contribute to harmonize the human mind between tolerance and self- restraint. All of these aims can only be expected to be achieved by appropriate education. The importance of education in the future is immense, and the timing is imminent.

In order for the future education not to be hated by subsequent generations, we need some techniques. The package-knowledge type teaching is evidently inadequate."[3]

---

[3] *Candid Science I*, 218–20.

ELENA GALPERN

The nature of the chemical bonds [of carbon cages] was a very intriguing question.

Elena Galpern. (courtesy of E. Galpern)

Elena Galpern[1] (1935–) was born in Moscow. She is a graduate of the Faculty of Physics and Mathematics of the Moscow V. B. Potemkin Pedagogical Institute. She spent her entire professional life at the A. N. Nesmeyanov Institute of Element-Organic Compounds (INEOS) of the former Soviet Academy of Sciences (now the Russian Academy of Sciences), from which she has now retired. Her research interests included the simulation of the structures of carbon clusters by quantum chemical methods. She was the first scientist in the world to arrive at the stable truncated icosahedral structure of C_{60} (subsequently named the *buckminsterfullerene*) by quantum chemical calculations. She and her mentor published their results in 1973 in the prestigious Russian periodical *Doklady Akademii Nauk* (Proceedings of the Soviet Academy of Sciences).[2] Nonetheless,

---

[1] I. Hargittai and M. Hargittai, *Candid Science I*, 322–31.
[2] D. A. Bochvar and E. G. Gal'pern, "Hypothetical Systems: Carbododecahedron, s-Icosahedron, and Carbo-*s*-Icosahedron" [in Russian]. *Doklady Akademii Nauk SSSR* 209 (1973): 610–12.

her work might have sunk into oblivion had it not been for the experimental observation and, later, the production of this molecule.[3] We recorded our telephone conversation with Galpern (in Russian) in 1994. We also exchanged written correspondence. Subsequently, we met with her in person, too.

Galpern told us about the background of the work that led to the publication of their 1973 paper reporting the stable C_{60} truncated icosahedral structure.

"At the end of the 1960s, we were working on a project that, at the beginning, did not have any direct connection to the modeling of carbon materials. Our research institute, INEOS, in Moscow, was busy with the chemistry of π-complexes of transition metals, and in particular, with ferrocene. The director of INEOS was Aleksandr Nikolaevich Nesmeyanov who was also the former president of the Soviet Academy of Sciences at the time. He had repeatedly told us all in the institute to look for the synthesis of new heteroorganic compounds in the form of endohedral polyhedral clusters M@C_nH_n with saturated carbon skeletons. One or a few heteroatoms would be included within the carbon cage.

The simplest of such systems might be envisaged by linking the two rings in ferrocene or dibenzenechromium or cyclophanes with polyene or polyyne groups. Thus, sewing the rings together would create the cage structure. The stability of such inclusion cage complexes and the nature of their chemical bonds were very intriguing questions. These questions constituted a research project of the Quantum Chemistry Laboratory of INEOS in the late 1960s.

At the time, however, there were great difficulties for computational work in chemistry in the Soviet Union. Therefore, we decided to restrict the computational work to the investigation of the stability of the all-carbon cage. This was considered to be the first step toward the investigation of the polyhedral inclusion complexes. The work was started with the C_{20} molecule of dodecahedral shape, called carbododecahedron. Since it was difficult to predict its stability in the valence approximation, we decided to use the Hückel method, which had proved to be successful for classical conjugated hydrocarbons...

The small size of the carbon cage was a limitation as it greatly restricted the choice of atoms to be placed within the cage. Thus, larger systems were sought for further calculations, and eventually the truncated icosahedron was selected. The corresponding carbon cluster of 60 atoms was called *carbo-s-icosahedron*. The C_{60} cluster was found to have a closed ρ-electron system, and a large enough energy gap to separate the occupied and vacant energy levels to provide kinetic stability. Besides, the size of C_{60} was sufficient for inclusion complexes of a wide range of atoms; the diameter of the inscribed sphere is 6.35 Å, assuming the C–C bond length to be 1.40 Å. Thus, the theoretical aspects of Nesmeyanov's question about the possibility of inclusion complexes

---

[3] H. W. Kroto, J. R. Heath, S. C. O'Brien, R. F. Curl, and R. E. Smalley, "C_{60}: Buckminsterfullerene." *Nature* 318 (1985): 162–163; W. Krätschmer, L. D. Lamb, K. Fostiropoulos, and D. R. Huffman, "Solid C_{60}: a new form of carbon." *Nature* 347 (1990), 354–58.

of polyhedral hydrocarbons were reduced to finding suitable models of new allotropic modifications of molecular carbon."[4]

We asked her what she was telling her friends or relatives about her discovery

"I have two kittens and a dog and when I try to talk to them about fullerenes, they stare at me with great bewilderment."[5]

---

[4] *Candid Science I*, 324–27.
[5] Ibid., 331.

DARLEANE C. HOFFMAN

We need to investigate the university climate and the concepts of tenure and why so many women choose not to even apply..

Darleane C. Hoffman in 2004 at Berkeley; she is wearing a pin for the element 106, Seaborgium. (photo by M. Hargittai)

Darleane C. Hoffman[1] (1926–) was born in Terril, Iowa. She received her PhD degree from Iowa State University in 1951. She has several positions, including senior advisor and charter director for the G. T. Seaborg Institute at the Lawrence Livermore National Laboratory. She has also held positions at Oak Ridge National Laboratory in Tennessee, the Los Alamos National Laboratory in New Mexico, and the Lawrence Berkeley National Laboratory in California. Her many distinctions include membership in the Norwegian Academy of Science and Letters (1990), the National Medal of Science (1997), and the Priestley Medal of the American Chemical Society (2000). We recorded our conversation in 2004 at her office at Berkeley.

---

[1] I. Hargittai and M. Hargittai, *Candid Science VI: More Conversations with Famous Scientists*, (London: Imperial College Press, 2000), 458–79.

Glenn T. Seaborg and Darleane Hoffman in the mid-1980s at Berkeley. (courtesy of D. C. Hoffman)

We asked Hoffman for her views on the question of the origin of elements in the universe, noting that the origin of the lighter elements is more or less understood but that it seemed to be a more difficult question for the heavier elements.

"The current theories postulate r-process nucleosynthesis, that is, rapid neutron capture until you finally get up to the highest reaches of the Periodic Table as we know it.[2] The highest atomic number element that we know of in nature is plutonium-244, which I and some of my coworkers found remnants of many years ago, in 1971. Of course, there is a lot of uranium in nature and it is formed by successive capture of neutrons by lighter elements. It was discovered in 1789 and is the heaviest element found in macroscopic quantities. This is how things were until about 1940. Then, when Ed McMillan and Phil Abelson were trying to investigate neutron-induced fission of uranium (just reported by Hahn & Strassmann in Germany) here in Berkeley, they discovered neptunium, which was the first synthetically produced transuranium element.

Your finding plutonium in 1971 is included in your National Medal of Science citation, 'discovery of primordial plutonium in Nature.'

Yes, that is among the things referred to in the Citation. Pu-244 is the longest lived isotope of plutonium, with a half-life of about 80 million years. Of course, it is a question, whether or not it was formed in the last nucleosynthesis of heavy elements in our

---

[2] The r-process nucleosynthesis—by rapid neutron capture—is supposed to occur in supernovae. This process creates about half of the neutron-rich nuclei of elements heavier than iron.

solar system. There is also the possibility that it might have accreted from extraterrestrial sources as our earth traveled through the galaxy. But we think that it is more likely primordial.[3]

This field is obviously a very competitive field—who finds and publishes a new element or isotope first. This then brings about the danger that you might rush with a publication before you can be absolutely sure of your findings.

Exactly. We also fell into that trap with our report of production and identification of the element 118 decay chain, except that in this case we were victims of fraud."

Fraud?

"Yes. It was perpetrated by one of our accomplished and trusted coworkers whom we brought here from the Gesellschaft für Schwerionenforschung in Darmstadt, Germany, to help build the instrument known as the Berkeley gas-filled spectrometer. We thought that we had observed three decay chains from element 118 produced in the reaction of lead-208 with krypton-85, and the results were published in 1999. Then we continued with additional experiments to try to do an excitation function for the reaction and weren't able to repeat it. We looked very carefully at the data tapes that had been saved and couldn't find evidence for the originally reported events on them. None.

Why did he or she do that?

If I could answer that I would be a psychiatrist instead of a nuclear scientist.[4]

Please, tell us something about Glenn Seaborg.

He was a wonderful resource. He would have a brown-bag lunch once a week, and he would remember everything because he wrote everything in his journal. More than that, every day he talked into a tape recorder, even when he was away, and when he came back his secretary would transcribe it. He started his diary when he was very young and kept doing it all his life. I wish I had done that! Many of his books were based on his journals. He had volumes of his journals. When I was here on sabbatical, in 1978–79, I remember the brown-bag lunches in his office, when people would tell what they were working on and we would talk about it in detail.

When I moved here as a professor, my husband told me that I should not be too surprised if I didn't get too many graduate students because maybe some of the young men would not want to work for a woman. I said, don't be absurd! Then when it was time to go down for the first party of incoming students, Glenn told me, let's go together and see who is there. I am very short, but he was very tall, about 6 ft. 6 in., and he would look around and introduce himself. Of course, at that time, all the students knew who he was—so I had no trouble attracting students—even if not for my own appeal! He was very helpful. And he continued to have his famous brown-bag lunches with my group. He traveled a lot. Helen, his wife, whom I also know very well, is a wonderful woman, too. She almost always went with him on his extensive travels and that must have made it easier for him. She accompanied him when he had to go to Washington on many occasions because he was the former chairman of the US Atomic Energy

---

[3] *Candid Science VI*, 460.
[4] Ibid., 475.

Commission (1961–1971) and subsequently was often asked to Washington, DC, to give his advice. Whenever I needed to find out about something, I just ran over to his office in the next building, and he often said, 'I can look this up in my journal to make sure.' He could tell me everything I had done, what I ate on a certain day, as well as everything else I wanted to know. He had a fantastic memory, which, of course, was reinforced by his faithful recording of his daily activities in his journal each day, a practice he began when he was twelve years old! He was also always extremely enthusiastic about students and he had a remarkable way of communicating with them."[5]

[5] Ibid., 471–72.

ROALD HOFFMANN

I have always been interested in instances in human history where the spirit and knowledge has come through, through periods of oppression and suppression, where it can't be expressed.

Roald Hoffmann with Istvan and Magdolna Hargittai in Reading, England, 1982.

Roald Hoffmann[1] (1937–) is a chemist and writer at Cornell University. He was born in Złoczow, Poland, now the Ukraine. He and his mother survived the Nazi occupation by hiding; his father was murdered by the Nazis. He graduated from Columbia University in 1958 and received his PhD degree at Harvard University in 1962 under the supervision of William N. Lipscomb. He interacted with Robert B. Woodward in working out what became known as the Woodward-Hoffmann rules of chemical reactions, based on the notion

---

[1] I. Hargittai and M. Hargittai, *Candid Science I*, 190–209.

of the conservation of orbital symmetry. In 1981, he was awarded the Nobel Prize in Chemistry jointly with Kenichi Fukui "for their theories, developed independently, concerning the course of chemical reactions."[2] We recorded our conversation in 1994 when Hoffmann and his wife, Eva, were visiting in Budapest.

On his schooling.

"My schooling was interfered with by the War [WWII]. First, there were a few months in a Ukrainian school in Złoczow. Then there was the second and third grade in a Catholic school in Krakow in Polish. My fourth grade was taught in Yiddish in a displaced persons refugee camp in Austria. Then a little bit in German in Germany, and eventually in the fifth and sixth grade, everything was taught in Hebrew in Munich. I learned algebra for the first time in Hebrew. All of this just to show my refugee background, with the mixture of languages that many children in the chaotic postwar period experienced. Then we succeeded in getting to the United States, of course, where all things went well. I attended New York City public schools, first a public school in Brooklyn, then a special science public school called Stuyvesant High School. There was a wonderful concentration of talent in that all-boys school in New York City. Then I went to Columbia for college. So I grew up in New York City. Eventually, I went to graduate school at Harvard.

We were poor; it was not easy for my parents to begin in a new country. So, in fact, we had no money to buy a book for me until I was sixteen. I can't remember any particularly inspiring teachers until high school. I remember, though, that I was not particularly interested in chemistry. At Stuyvesant High School there were advanced courses in every field. They would be called in the United States today 'advanced placement courses.' This was a kind of second course in each subject. I took such courses in biology and physics but not in chemistry. I also took a lot of courses in mathematics. At the end of high school I intended to go into medical research. That would have been a compromise between my parents' desire that I should become a doctor and what I wanted to do, which was some sort of scientific research. I started the university as a premedical student. But that did not last more than a year, and I somehow drifted into chemistry.

I had some unusually good teachers in high school and they were in mathematics and in biology. At Columbia University, the teachers I had in humanities courses were just wonderful. The ones in science courses were okay, but I don't think I hit one that inspired me until the last year of college. I really think that if I had not encountered those teachers in my last year— George Fraenkel and Ralph Halford—I would have gone into the humanities.

The humanities were so seductive, especially the history of art. I had fantastic teachers in Japanese literature, in the history of art, in English literature, poetry, and other

---

[2] "The Nobel Prize in Chemistry 1981," *Nobelprize.org*. Nobel Media AB 2013, http://www.nobelprize.org/nobel_prizes/chemistry/laureates/1981/.

literature courses. The world was just opening for me, and those were the most inspiring teachers I had."[3]

On his language skills.

"My first language was Polish. Ukrainian and Yiddish were also around and I knew them early on. Then came German, followed by Hebrew. Children learn quickly and forget quickly. By the time I came, at age eleven, to the United States, German was my dominant language. Then English took over. I was not very nice to my parents. I made them speak English to me.

Four languages were spoken interchangeably at home: Polish, Yiddish, English and German. Subsequently, I learned two other languages well, Russian and Swedish for one reason or another, and another language, French, not so well because I studied it only at school. English is, however, the only language in which I can write; it *is* my native language. Native speakers may detect a slight accent in it, and I have just a few small problems in the writing occasionally, such as confusing 'like' and 'as' and 'that' and 'which.' Occasionally my constructions are a little funny, but English is my language and I love it.

I think more important to writing may be being an outsider. Knowing several languages puts you a little outside the one language you have been thinking in. You ponder of things more than the native speakers. Being an outsider comes from being an immigrant in a society. And Jews have been outsiders in other ways. I have also switched subfields of chemistry; I felt like an outsider when I started in organic and in inorganic chemistry. I sort of like that feeling, for you get a different perspective. Coming from the outside, at first it's a little dangerous and difficult but I like the feeling of penetrating the walls built of jargon and custom around a field."[4]

---

[3] *Candid Science I*, 192–93.
[4] Ibid., 205.

ISABELLA L. KARLE

Efforts are being made to improve the teaching of science at all levels.

Isabella L. Karle in 2000 in Washington, DC. (photo by M. Hargittai)

Isabella L. Karle[1] (1921–) was born in Detroit, Michigan. She received her master of science degree in 1942 and her PhD degree in 1944 from the University of Michigan at Ann Arbor. She participated in the Manhattan Project at the University of Chicago. She joined the Naval Research Laboratory in 1946 and remained affiliated with it for the rest of her career. She is a member of the US National Academy of Sciences, received the National Medal of Science, the Aminoff Prize from the Royal Swedish Academy of Sciences, and numerous other awards. She was married to Jerome Karle (see next entry). We recorded our conversation in 2000 at the Naval Research Laboratory in Washington, DC.

---

[1] I. Hargittai and M. Hargittai, *Candid Science VI*, 402–21.

Karle in the early 1950s at the University of Michigan in Ann Arbor with a gas-phase electron diffraction apparatus. (courtesy of I. Karle)

We asked whether and to what extend Isabella and Jerome Karle delineated their scientific research activities.

"Jerome and I work together separately. We both did our graduate degree work with the same man, Lawrence Brockway, in gas electron diffraction. Then we spent some time on the Manhattan Project at the University of Chicago in the chemistry laboratory. Our activities differed from those in our PhD programs. I was making plutonium chloride from a crude plutonium oxide, and Jerome was trying to make the pure metal. The objective of the Chicago project was to make pure plutonium without any impurities. A number of different paths were being tried simultaneously. I don't know how the plutonium metal was produced eventually, but Jerome succeeded in making some rather pure metal directly from the oxide, and I managed to make pure plutonium chloride by many different paths using very high temperatures in a vapor phase process. The resulting crystals were absolutely beautiful. They were jade green and they grew with large, smooth faces.

After we finished our particular projects, we could have stayed in Chicago, but there was no particular urgency since the thrust of the project had moved to Los Alamos."[2]

---

[2] Ibid., 406.

On managing to have children (the Karles have three daughters) and staying at the top of their science at the same time.

"We were fortunate that after World War II, there were many women of grandmother age whose children didn't want to live on the farm any more. The mountains of Virginia are about sixty miles from here. Many of the younger people came to Washington during the war and didn't want to return to the country after the war ended. The elder ladies came also and often served as live-in housekeeper/babysitters during the week. They would visit with their own children during the weekends. This procedure worked out very well for us until our children were old enough that we didn't need constant help any more.

Having children required organization, but I never felt them to be an obstacle to my career. It never interfered with our professional lives. When they got a little older, and by a little older I mean at least seven years old, we always took our children with us for our summer travels to various meetings in Europe, Japan, and the United States.[3]

There seems to be a gap of top science in the United States, leading the world, and the general ignorance of much of the population.

The National Academy of Sciences has had a very active committee for the last ten years or so, in trying to improve science education from kindergarten through twelfth grade. They have done quite a number of things to make it easier for schools that do not have teachers trained in science to be able to teach their students, by making sure that there are textbooks that even teachers who have no idea about science can use with their students. I find with my grandchildren that there has been quite a bit of science introduced from the very beginning. I don't know how many schools they reach. This is a very varied country. There are rural schools and small towns that are poor and don't have the facilities and money for hiring good teachers. But efforts are being made to improve the teaching of science at all levels."[4]

---

[3] Ibid., 409–10.
[4] Ibid., 419.

JEROME KARLE

Ethics is very important in science...this is my message.

Jerome Karle in 2000 in Washington, DC. (photo by I. Hargittai)

Jerome Karle[1] (1918–2013) was born in Coney Island, in Brooklyn, New York. He received his master of arts degree in biology from Harvard University in 1937, and his master of science degree in 1941 and his PhD in physical chemistry in 1943 at the University of Michigan. He participated in the Manhattan Project at the University of Chicago. He became affiliated with the Naval Research Laboratory in 1946 and stayed with it for his entire career. Karle and Herbert A. Hauptman received the 1985 Nobel Prize in Chemistry "for their outstanding achievements in the development of direct methods for the determination of crystal structures."[2] We recorded our conversation in 2003 in Jackson, Mississippi, during the Conference on Current Trends in Computational Chemistry.

---

[1] I. Hargittai and M. Hargittai, *Candid Science VI*, 422–37.
[2] "The Nobel Prize in Chemistry 1985," *Nobelprize.org*. Nobel Media AB 2013, http://www.nobelprize.org/nobel_prizes/chemistry/laureates/1985/.

The Karles and the Hargittais in 1978 in Pécs, Hungary.

About the work leading to the direct methods in X-ray crystallography.

"It was done stepwise. Herb [Hauptman] and I had derived the mathematical relations that were necessary for phase determination by about 1955–56. However, for practical application an additional step was needed. We had to assign a place in the crystal that could be referred to as the origin. This step needed different considerations for each of the 230 three-dimensional space groups and meant a lot of work for Herb and me. Not all the specifications of the origin are different for all space groups but each one had to be considered one after the other. The great virtue in specifying the origin was that it allowed the determination of the phases directly from the intensities. Herb left our laboratory about 1960.

Isabella and I worked out a modus operandi for phase determination, which we called the symbolic addition procedure. The first crystallographic structure that Isabella and I worked on alone was a centrosymmetric one that had been attempted at several laboratories. It concerned the cyclohexaglycyl molecule. There were four conformational isomers in the unit cell. This structure established for the first time the coexistence of conformational isomers in peptides and also reliable parameters for the beta-hairpin turn and its hydrogen bond. This paper was one of the most quoted papers in the *Citation Index.* "[3]

---

[3] *Candid Science VI*, 430.

About the greatest challenge in his life.

"Getting into a graduate school that would accept me. There was this young person who had this wish, who worked hard, and who was denied the privilege."[4]

He was not eager, though, to talk about this.

"I went to Abraham Lincoln High School and to City College of New York from which I graduated when I was nineteen years old. I could have graduated when I was eighteen years old, but I felt that I was too young and spent one-half year more than was required in high school and college.

What happened then?

I didn't know much about graduate schools until my last year in college. It was virtually impossible to get into medical school for reasons that are not very complimentary for society at that time.

Were there actual rules against Jews or it just happened?

There were no rules. I don't know whether I should speak about this.

Please do.

I went to Harvard and spent a year there obtaining a master's degree in biology. I had the illusion that being a good student was all that was necessary to get admitted to medical school. I applied to Harvard and some other places and, of course, I was turned down. I wanted to try again early the next year, and I was allowed to have a conversation with the dean of the medical school. The only thing I got from him was a harangue. He said, 'We have enough Jews in Massachusetts, we don't need any from New York City.' He was not at all interested in my record as a student. For example, when I graduated from City College, I received the first award given at graduation for 'excellence in the natural sciences.'

What did you do then?

I had applied to various graduate schools just to do graduate work and I was turned down by all of them. So I wasn't doing anything. Then, there was just a stroke of luck. In the summer of 1938, I was working in Coney Island and a good friend of mine, with whom I still communicate, told me that exams were forthcoming for civil service jobs in the New York State Health Department. I took the exam and I had the highest grade among those they accepted. There was a rule that after a certain period, perhaps three months, the Health Department could not dismiss anyone without an explanation. I stayed for about two years. I learned only later that they had wanted me to leave, along with the rest of the people who arrived when I did, but my boss said that if they tried to dismiss me, he would not accept that and that he needed me for his work. This fine gentleman's name was F. Wellington Gilcrease. He remained my good friend, and we kept in touch until the end of his life. I did not know that he saved me from dismissal until I left to go to the University of Michigan. During those two years,

---

[4] Ibid., 435.

I was saving up money as I knew I couldn't get any money from graduate school. At that time, someone told me that if I went to the University of Michigan, I would be treated properly. That is why I went there. After the first year, I was funded to continue my education at the University of Michigan. After that, I have never experienced any anti-Semitism."[5]

---

[5] Ibid., 425–26.

REIKO KURODA

Kuroda told us that the enormity of her project and her grant was "killing" her.

Reiko Kuroda in 1999 in Budapest. (photo by I. Hargittai)

Reiko Kuroda[1] (1947–) was born in Akita Prefecture, Japan. Until recently, she was at the Department of Life Sciences of Tokyo University. She was the first female full professor in the history of this most prestigious institution. In 2012, she retired, because she had reached the mandatory retirement age, and started a new laboratory at the Tokyo University of Science. She has been a much decorated scientist; the latest of her distinctions was in 2013, the L'Oreal-UNESCO award Women in Science. We recorded our conversation with Kuroda in Budapest in 1999.

---

[1] I. Hargittai and M. Hargittai, *Candid Science III*, 466–71.

Kuroda in 1996 in Tokyo with, clockwise from her left, Elizabeth Watson, James D. Watson, Francis Crick, Odile Crick, Lise Jacob, François Jacob, and Susumu Tonegawa. (courtesy of R. Kuroda)

We discussed with Kuroda such sensitive subjects as, for example, how they teach the history of World War II in Japanese schools. She found it more convenient that we present our conversation in a narrative form, rather than quoting her directly.

Reiko Kuroda graduated from the famous women's college Ochanomizu University in Tokyo in 1970 and got her master of science and PhD degrees from Tokyo University in 1972 and 1975, respectively. Her research was in X-ray crystallography. For postdoctoral work, she went to King's College in London. At the time, it was difficult to find a job in Japan, especially for a woman with a PhD degree. At that time, the job market in Japan was not an open system. Rather, a professor would just mention to a colleague that he had a good student available. But Reiko Kuroda did not get this help either, since it was difficult to find jobs, even for men. She was given the advice that the best thing for women was to get married, and she was offered help in finding a husband. She declined the offer.

Kuroda was attracted by a postdoctoral position with Stephen Mason in the Department of Chemistry at King's College in London. Her English was not very good when she arrived in London, but she was determined to improve it. She got involved in teaching, and that helped too. The Nobel laureate Maurice Wilkins asked her to take over teaching his course on macromolecular assemblies.

In 1981, Kuroda accompanied Wilkins on a trip to Japan. He was a little worried about the trip. He had been involved with the Manhattan Project during World War II, as one of the British team working in California. Of course, his Japanese hosts knew about his work on the bomb. Wilkins asked Reiko, 'What should I talk about?' He didn't want to upset anybody.

Kuroda thinks that all Japanese condemn the dropping of the atomic bombs. She doesn't think that these bombs made much difference in ending the war because by then, the situation in Japan had reached a point that would have made it impossible to continue the war. There was a shortage of food, a shortage of everything. To the question of why the Japanese never mention the Japanese atrocities in China and Korea and elsewhere, Kuroda responded that it depended on the way the question was put to the Japanese people. When they are asked about the bombs, they won't engage in discussing other things. She had never learned in school about what had happened during the war. She found out about many events later, from newspapers, and that was when she learned that other countries condemned the atrocities committed by the Japanese military during the war. It helped that she could read English. History class in her school started with the beginnings of civilization, and by the end of the academic year, they had not even reached the Meji period, which started in 1868. This was a recurring pattern. They would not have exam questions on the time after the Meji restoration. She remembers that they always had to finish history courses in a rush lest the school year end without their completing the curriculum.

Eventually, Kuroda became Honorary Lecturer and finally a research fellow and won a permanent position in a tough competition in London. By then, she was in the Biomolecular Structure Unit at the Institute of Cancer Research. She continued her teaching at King's College as Visiting Lecturer.

For the first several years in London, Kuroda was thinking of going back to Japan. She had been warned that if she stayed away for more than two years, she would lose an important feeling of harmony and would become too westernized, and then it would become impossible to get accepted again by Japanese society. After two years, Kuroda got a job offer from Japan to teach in a private finishing school for girls. The position was only for two years, and there was no chemistry department there; they just wanted her to teach some science and to teach English. She stayed in London.

She felt settled, but she was also intrigued to try her hand in the Japanese system and, after several years, she decided to submit an application for an opening at Tokyo University. To her great surprise, she landed one of four associate professor positions from among 140 candidates, and in 1986, she became the first female associate professor in natural sciences on her campus. When she was appointed full professor in 1992, she was the first female full professor in the history of Tokyo University. Some worried that she might not be able to understand the culture, the harmony, but there was no problem.

When her colleagues go out in the evening for a drink, sometimes they ask her to join them. Normally, she is so involved with her experiments that she is not eager to go out anyway. When her colleagues have foreign visitors, they like her to join them because of her fluency in English. Her good English may be a mixed blessing, though. She had been warned not to speak English in front of her colleagues. She had read articles about students returning to Japan after spending years abroad who were advised not to use the proper pronunciation in an English class but adopt instead the Japanese pronunciation lest the Japanese teacher of English become unhappy....

The rest of Kuroda's career in Japan was a great success. She received enormous grants for her cleverly titled project, 'Chiral Morphology'. She was also given tremendous responsibilities, and her career appeared to be a showcase example of the changing climate for female scientists in Japan. Thanks to her endurance and will power, she withstood the trials of her situation and took her challenges in stride. When asked if she could have one wish what it would be, she hesitated for a long time before answering. First, she said, she wanted to make sure that the wish could be anything, however unrealistic. Finally, she said, it would be nice to have a family and continue her work as well.

YUAN TSEH LEE

I believe in one thing. I believe in education.

Yuan Tseh Lee in 2005 in Lindau, Germany. (photo by I. Hargittai)

Yuan Tseh Lee[1] (1936–) was born in Hsinchu, Taiwan. He is a physical chemist who received his bachelor of science and master of science degrees from National Taiwan University in 1959 and 1961, respectively. He continued his graduate studies at the University of California, Berkeley, and received his PhD degree in 1965. Following postdoctoral studies at Berkeley and Harvard, he taught at the University of Chicago, and then in 1974 returned to Berkeley. He moved back to his native Taiwan in 1994. Lee shared the 1986 Nobel Prize in Chemistry with Dudley R. Herschbach and John C. Polanyi "for their contributions concerning the dynamics of chemical elementary processes."[2] We recorded our conversation in 2005 in Lindau, Germany.

---

[1] I. Hargittai and M. Hargittai, *Candid Science VI*, 438–57.
[2] "The Nobel Prize in Chemistry 1986," *Nobelprize.org*. Nobel Media AB 2013, http://www.nobelprize.org/nobel_prizes/chemistry/laureates/1986/.

Lee in 2001 at the centennial Nobel Prize award ceremony. Lee is first at the lower left; César Milstein is in front of him, partly hidden; Marshall W. Nirenberg is on Lee's left; Jerome Karle, partly hidden, is behind Lee; and Richard Ernst is behind Karle, to the left. (photo by M. Hargittai)

We remarked to Lee that for an outsider, living in Taiwan must be pregnant with various tensions.

"There has been such a tension, but it has taken a different turn. The Taiwanese were oppressed by the Japanese for fifty years during colonization. After the Second World War we waived the flag and we were so happy to return to the Motherland. But then, the regime that came to Taiwan under Chiang Kai-shek was so corrupt that people felt that it was worse than the Japanese. The difference was that the Japanese ruled by law. It was not good for the Taiwanese people, but it was calculable and we knew how we could get by. When the Mainland Chinese came, they were corrupt. The former Japanese properties became national properties but the Mainland Chinese officials expropriated those properties. The young socialists were especially upset. They demanded a fair society; they were not card-carrying party members, but people hoped that they would help to remove the Chiang Kai-shek regime. Instead, the Mainland Chinese said that Taiwan could keep its government and military and police so long as you come back to China. Thus people realized that the communist party served the oppressors rather than the people. The people in Taiwan felt betrayed. This is why so many would like to have independence declared. I don't think though that independence is so important. The people in Taiwan need and seek fair treatment.[3]

---

[3] *Candid Science VI*, 449.

At some point, the United States changed its policy toward Taiwan.

Taiwan is so small compared with Mainland China that Taiwan would not survive without US protection, and that has not changed. In this sense, we are helpless. Politics is a very pragmatic activity. China is a very large country and the trade between the People's Republic and the United States has become very important. There is no real choice for any country in the world when the question comes up about choosing which country to recognize, Mainland China or Taiwan. Everybody says that we are a democratic society and they would like to be our friends, but Mainland China forces them to choose between them and Taiwan and they just can't afford not to choose Mainland China. Only a few small countries in Africa and in the Pacific region still recognize Taiwan. None of the G8 and other developed nations recognizes Taiwan anymore. What we hope for in Taiwan is that Mainland China one day will become democratic and then unification won't be a problem.

Are things evolving in that direction?

They have to. China faces great difficulties at the present time. There are two problems. One is corruption in China. Socialism and communism, which held people together for the last fifty years have gone bankrupt. People again face a dilemma. KMT or Kuomintang—Chiang Kai-shek's party—was so corrupt that the Chinese threw them out from China to Taiwan. They are still corrupt in Taiwan. They say that the communist party in China is even more corrupt than Kuomintang was a long time ago. The second problem is that before the socialist revolution there was a large gap between the rich and the poor and the young people resonated with the socialist ideas. They wanted the society to become egalitarian and fair. Looking at the distribution of wealth now, they see an even wider gap than there was before the revolution, so the purpose of the socialist revolution was lost. All the sacrifices appear to have been meaningless, in vain. The situation in Mainland China may develop in a dangerous direction. People no longer believe in the socialist revolution, and the country is moving in the direction of so-called nationalism and patriotism. The recent anti-Japanese movement is part of this development. Of course, Japan should think about this too; they have never done repentance in the proper way for the crimes they committed during the Second World War. Germany has done such repentance and promised that she would never do it again. Japan has given a lot of money to China but never expressed their condemnation of the war crimes in a proper way. Sadly, however, China seems to need to use the anti-Japanese sentiments to unify people and this is dangerous."[4]

---

[4] Ibid., 449–50.

JEAN-MARIE LEHN

I tried to contradict Freud, but he always won.

Jean-Marie Lehn in 1995 in Budapest. (photo by I. Hargittai)

Jean-Marie Lehn[1] (1939–) was born in Rosheim, France. In 1987, he shared the Nobel Prize in Chemistry with Donald J. Cram and Charles J. Pedersen "for their development and use of molecules with structure-specific interactions with high sensitivity."[2] Lehn has had a double appointment at the Institut Le Bel, Université Louis Pasteur in Strasbourg and at the Collège de France in Paris. He has also had high international visibility, especially in Europe, having been active in representing chemists and chemistry on social issues in addition to issues directly concerning science. We recorded our conversation in 1995 in Budapest.

---

[1] I. Hargittai and M. Hargittai, *Candid Science III*, 198–205.
[2] "The Nobel Prize in Chemistry 1987," *Nobelprize.org*. Nobel Media AB 2013, http://www.nobelprize.org/nobel_prizes/chemistry/laureates/1987/.

About how Lehn's supramolecular chemistry differs from traditional molecular chemistry.

"The main difference is that I was looking for the more general concept. It's quite clear that chemists have been developing molecular chemistry for many years. By making and breaking bonds, you can create molecules. This field has by now reached fantastic complexity. Chemists are extremely knowledgeable in doing that, with very high precision and efficiency. Not everything is known yet, of course, and there is still a lot of things to be done in synthetic chemistry. Molecular chemistry has a fantastic future. However, once you've reached a certain level of development, at which you can make and break covalent bonds to build molecules almost at will, the question you may ask is, can you do the same at the level of weak interactions? That is, at the level of intermolecular weak bonding. There are forces between molecules, which can bind them together weakly. The question is how can you manipulate these intermolecular interactions to generate supramolecular structures?

I consider supramolecular chemistry as the extension of molecular chemistry to the more complex systems. The task is to enable the chemist to handle non-covalent structures as he or she had learned to handle covalent structures. Because the bonds are weak, a number of new properties appear. I like to stress the analogy with information science. The reading of information is subtle and must be reversible. There is need for reading, and this requires weak interactions, in which you can bind and dissociate again. Also, supramolecular structures have a property which you may call healing. When they associate wrongly, they can dissociate and recombine. In other words, they have a self-healing mechanism. One major characteristics of supramolecular structures is this capability for self-processes. There are processes that occur spontaneously, but they don't occur for covalent bonds at normal temperatures and pressures because they are too stable for that. You may in this respect consider molecular chemistry as stable chemistry or fixed-structure chemistry, and supramolecular chemistry as fluid chemistry. The term fluid here means that the structures can be arranged and rearranged, assembled and disassembled, depending on the surroundings. Adaptation is another important characteristics of supramolecular systems. A molecular structure, as far as connectivity and gross change is concerned, is very stable, not influenced so much by the medium. A supramolecular structure can, in principle, adapt to the medium."[3]

About coping with increasing responsibilities.

"One learns with time. Before accepting various functions, especially those that take me away from chemistry, I have to think hard to convince myself to do it. One has to be careful accepting things in which one may be less and less competent. On the other hand, as you think more about what you are doing and about the place science occupies in society, you feel more commitments to many things, to your fellow scientists and to science itself. Science is the strongest force in transforming society. The progress of mankind depends on knowledge, that is, on science and its applications. By saying

---

[3] *Candid Science III*, 201–2.

this, I don't mean that, for instance, philosophers and composers are useless, not at all! I only mean that the objective transformation of society is based on the knowledge we have gained about nature. We need not be arrogant scientists. At the same time, just because we may be accused of 'scientism,' we should not shy away from stressing the importance of science. Science has nothing to do with any dogma. Science ceases to exist when there is a dogma. For me science is an approach, a way of taking things in a rational fashion, thinking, and trying to do the best with what we know, trying to get solutions in complicated cases to the best approximation, knowing that this is not the perfect solution, and that there is no absolute solution. This is just the antithesis to many very careless and emotional approaches that we see in the world around us these days. Science can merely offer some options for rational solutions, and it's then up to the decision-makers whether they want to use these options or not. Of course, one has to be careful about the deeds of these so-called decision-makers or politicians. I'd only cite what is happening in the south of Europe. I am sure that in many places at war the people would be quite happy to live together. However, because of political ambitions, artificial problems are created in cases where they did not exist. Then, of course, when you raise antagonism, people finally think that they do in fact have enemies. The people involved often didn't even know, or did not care, that their present enemies were of other race or culture. They only were made to be aware of this. That's crazy. I believe that the scientific spirit is the antidote of this."[4]

On his youthful reading experience.

"When I read [Sigmund] Freud for the first time, it was a big shock to me. When you're brought up in a conventional way, it's something which makes you think very hard. I remember that I tried to contradict Freud, but he always won."[5]

---

[4] Ibid., 202–3.
[5] Ibid., 205.

WILLIAM N. LIPSCOMB

People who make discoveries do not use the aesthetic side until afterward.

William N. Lipscomb with his Nobel diploma in 1995 at Harvard University.
(photo by I. Hargittai)

William N. Lipscomb[1] (1919–2011) was born in Cleveland, Ohio, and grew up in Kentucky. He attended Picadome High School (today, Lafayette High School) in Lexington, Kentucky. He earned his bachelor of science degree from the University of Kentucky and did his PhD studies at the California Institute of Technology. At the time of our encounter, he was the Abbot and James Lawrence Professor Emeritus of Chemistry at Harvard University. He

[1] I. Hargittai and M. Hargittai, *Candid Science III*, 18–27.

received the Nobel Prize in Chemistry 1976 "for his studies on the structure of boranes illuminating problems of chemical bonding."[2] Lipscomb put great emphasis on the outlook of his publications. His beautiful hand-drawn borane structural formulae became his trademark. Lipscomb was a recognized performing clarinetist. We recorded our conversation in 1995 in Lipscomb's office at Harvard University.

About his early development.

"I grew up in Kentucky. My father was a physician, an MD, and my mother a music teacher. She taught voice. When I was eleven years old, she gave me a chemistry set, one of those things you buy in a store, and I started experimenting with it. That's not unusual. Many boys get these sets and play with them for two weeks. Only I didn't stop playing. Then I discovered that I could buy additional apparatus and chemicals using my father's privilege at the drug store. He had a special rate because he was a physician. So I began building up a home laboratory. I did both chemical experiments and electrical experiments. I also put a wire across the street to some friends who were also interested in science. In fact, within about four houses either way there were several boys about my age. Among this group of boys, there were eventually two PhDs in physics, two MDs, one civil engineer, and me. We had quite a group interested in science. This is very important....

...The high school that I went to actually had very poor facilities. It was in the county, not the city, on the edge of Lexington. It was Picadome High School, and it eventually became Lafayette High School. They had physics and they had biology courses, but they didn't have a chemistry course. As I learned later, my father went to the superintendent of the county schools and said, you'd better have a chemistry class there, or I'll take my son out and put him downtown in the big high school. So they made a chemistry course, and I helped make it, actually. When I came to the course, the teacher said, well you already know all the chemistry we are going to have, why don't you just show up for examinations. So I stayed in the back of the room and did some experiments, a little piece of original research, but I didn't publish it. I still have the manuscript.

...Then I went to the University of Kentucky, which was not a very good university by the standards of the Big Ten. As I went through college on a music scholarship, I had a separate program of my own, of reading and trying to work out things that were not taught. For example, nobody in the chemistry department knew anything about quantum mechanics, and I studied it on my own. One of the physics professors helped me a little bit. I also studied mathematics with a man named Fritz John who went on to the Courant Institute in New York and eventually became a member of the National Academy of Sciences. He was a very great mathematician, but they didn't know that then. He had just come over from Germany because of

---

[2] "The Nobel Prize in Chemistry 1976," *Nobelprize.org*. Nobel Media AB 2013, http://www.nobelprize.org/nobel_prizes/chemistry/laureates/1976/.

Hitler's deportation and killing of the Jews. He escaped and was at an early stage of his research. I learned a lot of new mathematics from him, group theory, matrices, vectors, and so on. And I taught him about Maxwell's equations when I was an undergraduate because he wanted to know something about physics. He was a very important individual in my life.

I had a pretty rigorous training from my school. Then, after graduation, I had some offers, one to go to Michigan at $150 a month without any duties except to go to graduate school, and another one from Caltech at $20 a month to assist in their physics lectures. I went to Caltech to study physics, and I took the low-paying offer because I wanted to be in a good school. During my first year, I heard some lectures by Linus Pauling and switched to chemistry. I did my thesis with him and some people who worked with him. That was a big part of my beginning."[3]

Lipscomb's Nobel Prize was for borane structures, which he understood by applying the concept of the so-called three-center bond in which three atoms share a pair of electrons. To the chagrin of many chemists, Gilbert Lewis never received the Nobel Prize. Lewis is credited with the discovery of the covalent bond (by analogy, we might call it the two-center bond), which is far more widespread than the three-center bond. We asked Lipscomb what he thought about Lewis's missing Nobel recognition.

"I think the problem was that nobody understood the two-electron bond until Heitler and London's research in 1926. Then Hund and Mulliken developed the molecular orbital method. It's true that Lewis lived much beyond 1926, but Lewis himself did not understand the nature of the electron-pair bond in his early research. Failing that, I think his time sort of passed. It may have been a good idea to give him the prize but I can understand why they didn't, because Lewis himself certainly didn't formulate the physical basis of the bond."[4]

---

[3] *Candid Science III*, 19–21.
[4] Ibid., 23.

STEPHEN MASON

The natural sciences do not flourish in periods of zealous and effective authority.

Stephen Mason in 2000 in Cambridge, England. (photo by I. Hargittai)

Stephen Mason[1] (1923–2007) was born in Leicester, England. He was a science historian; Professor Emeritus of King's College, University of London; and Fellow of the Royal Society, London. His most famous work was *A History of the Sciences*, whose latest revision he left uncompleted. His last book was *Chemical Evolution*.[2] We recorded our conversation in 2000 in Cambridge, England.

Our extensive conversation was about the lessons of the history of the sciences and here we quote his response to our question: "What can we learn about the relationship between science and authority in the history of Science?"

"Broadly, we learn that the natural sciences do not flourish in periods of zealous and effective authority, particularly in cases where socially dominant institutions fear that

---

[1] I. Hargittai and M. Hargittai, *Candid Science III*, 472–95.
[2] Stephen F. Mason, *Chemical Evolution: Origin of the Elements, Molecules and Living Systems* (New York: Oxford University Press, 1991).

scientific innovations may jeopardize their power, as in the case of the Roman Church during the sixteenth and seventeenth centuries. The Protestant Churches were more divided, and political power lay primarily in the hands of local princes, some pursuing scientific studies themselves, such as Landgrave Wilhelm IV of Hesse (1532–92), who built his own astronomical observatory at Kassel. The king of Lutheran Denmark, Frederick II, patronized the observational astronomy of one of his nobles, Tycho Brahe (1546–1601), whose death left his assistant Kepler with astronomical data, measured over twenty years, and the post of Imperial Mathematician to emperor Rudolf II at Prague. In contrast, the observatory at Istanbul was destroyed, after only six years of operation (1575–80), by the command of Ottoman sultan Murad III (despite his astrological interests) following the ruling (*fatwa*) of his Muslim juriconsults that astronomical prying into celestial affairs had caused the 1580 outbreak of bubonic plague in Turkey.

By the late sixteenth century, no religious authority in Europe could exercise such a control over science policy, and, indeed, the Jesuits built observatories in order to compete with laymen in astronomy. Princes had become more and more resistant to the attempts of the Church to extend temporal power ever since the forgery in Rome of the Donation of Constantine during Carolingian times. It purported to give the Bishop of Rome the temporal powers of the Western Roman Emperor, and to delegate those powers to a chosen nominee, crowned as the Holy Roman Emperor, beginning with Charlemagne (c.742–814). These emperors had their own policies, and the Hapsburg emperor Rudolf II (1576–1612) collected to his Prague court notable scientists whatever their religion, appointing as Imperial Mathematician the Lutherans Tycho Brahe and then Kepler. During the sixteenth century, many princes strove to centralize their realms, taking over the Church within their territories as an instrument of state. During the seventeenth century the princes and the learned among their professionals and gentry came to appreciate the potential usefulness of the natural sciences, as illustrated by the writings of Francis Bacon (1561–1626), and they founded national observatories and academies of science. While the science academies of Italy were ephemeral, and natural science was dormant in Italy for a century or so after the time of Galileo and his immediate disciples, the scientific societies of England and France were enduring, and they served as models for the eighteenth century science academies of North America, Russia, Germany, and elsewhere in Europe.

Thereafter, the relationship between science and authority became more a question of political conformity rather than religious dissent. Joseph Priestley was ostracized in London much more for his sympathy with the French revolutionaries than for his Unitarianism, and John Dalton encountered disdain in London more for his association with industrial Manchester than for his Quaker beliefs. Prelates could be stirred to angry opposition by discoveries apparently at odds with their theological doctrines, such as, *On the Origin of Species by Means of Natural Selection, or the Preservation of Favoured Races in the Struggle for Life* (1859) by Charles Darwin (1809–82). But theology was no longer widely regarded as the queen of the sciences, nor did prelates retain any more the social standing and power they enjoyed in Galileo's time.

During the twentieth century, ideological factors were added to political conformity in the relationships between science and authority. In Germany, 1933–45, scientists, among others with traceable Jewish antecedents, were expropriated and exiled or extinguished, with an even more efficient ruthlessness than the Spanish monarchs and their Inquisition during the fifteenth and sixteenth centuries. The wealth sequestered from the Jews expelled from Spain in 1492 financed the discovery of the New World by the expedition of Columbus in search of a westerly route to Asia. The Spanish Inquisition pursued the *conversos,* descendants of Jewish converts to Christianity, and the *moriscos,* the corresponding Muslim converts, with vigor, and 45 percent of the cases recorded by the Court of the Inquisition of Toledo between 1575 and 1610 refer to these two categories, which included around 10 percent physicians and surgeons. But the Iberian monarchs and their grandees retained the services of *converso* physicians, reputed for their medical skills, just as Field Marshal Goering in Germany retained the services of the 1931 Nobel laureate biochemist, Otto Warburg (1883–1970), despite his Jewish antecedents, since Warburg was reputed to have a cure for cancer, from which Goering feared he might suffer.

Historically, such pogroms result in an enormous loss of talent, not only for the countries of the perpetrators, but also for humankind at large, and ultimately they fail in the quest for an extension of power over the natural and human world. The same holds for the lesser and more secular analogues after WW II, McCarthyism and Lysenkoism. The crusade of Lysenko (1898–1976) in the USSR against Mendelian genetics over the years 1948–65 came to an end when it became apparent that his version of Lamarckism, the inheritance of acquired characters, failed both in agricultural practice and in laboratory test experiments. Lysenko's disciples, whom he nominated as candidates for election to the USSR Academy of Sciences, were rejected by the physicists on the grounds that these candidates, while excelling in the rhetoric of criticism, produced no authentic and tested innovations. The rise and fall of Lysenkoism provided an object lesson for those sociologists who regard all scientific theories as wholly socially constructed ideologies, and who make no reference to the success or failure of the theories in the laboratory or the field."[3]

[3] Ibid., 486–88.

BRUCE MERRIFIELD

Don't ever let your kids have X-ray treatment unless it's absolutely critical.

Bruce Merrifield in 1996 at Rockefeller University. (photo by I. Hargittai)

Bruce Merrifield[1] (1921–2006) received his education and degrees from the University of California, Los Angeles. He was with the former Rockefeller Institute (now Rockefeller University) in New York City from 1949 to the end of his life. He invented a new methodology for the synthesis of peptides and small proteins. In 1984, Merrifield received the Nobel Prize in Chemistry "for his development of methodology for chemical synthesis on a solid matrix."[2] We recorded our conversation in 1996 in his office at Rockefeller University.

It took a few years for Merrifield to develop the methodology, and for this, he happened to have an ideal supervisor in Wayne Wooley, who let Merrifield work on his own idea and did not press him to publish right away. Rockefeller University was similarly an ideal

---

[1] I. Hargittai and M. Hargittai, *Candid Science III*, 206–19.
[2] "The Nobel Prize in Chemistry 1984," *Nobelprize.org*. Nobel Media AB 2013, http://www.nobelprize.org/nobel_prizes/chemistry/laureates/1984/.

Wayne Wooley. (courtesy of B. Merrifield)

venue for Merrifield's research, because he could devote himself to his project and not worry about its funding. Here is how Merrifield spoke about Wooley and about how the idea on solid matrix was developed.

"He was a fine biochemist, a brilliant man. He'd lost his eyesight in his mid-twenties. It was diabetes. But he carried on and maintained his lab, with a technician, of course. He only had to read something once—it was read to him—and he'd remember everything, and he could correlate information from many areas into a single idea.

I worked with him for several years on his projects, until I got the idea of solid phase synthesis. I told him the idea. It was kind of funny; we were riding up the elevator and he got off without a word. The next day he came in to see me and said, 'That may be a good idea, why don't you work on it?' That made all the difference. He could have said, no, that's not what I want to do, I don't want you to work on it. The Rockefeller Institute, later Rockefeller University, was set up more like the European system. You had a lab head, and people worked under him. So when I was in his lab, I worked on what he was doing. He directed it all. Then, after a while, he'd let you do a little bit on your own. That was standard; there wasn't anything unusual about it....

I was working on a peptide growth factor with Dr. Woolley, and we isolated it, and I sequenced it and synthesized it by standard methods. It was only a simple peptide of five amino acids, but it took me ten or twelve months before I could make it and a couple of its analogues. I realized that peptide synthesis was quite hard to do. People who had a lot of experience could have done it faster than that, but it was still slow business, and the yields were not very good. So I had in the back of my mind that there must be a better way to do it. Then one day or one night, I don't know which, I just had this idea: couldn't we anchor the peptide onto a solid support? If you could put the first amino acid onto a substance that was insoluble in all the solvents you used, and then you added the next amino acid and the next one and the next one, you would build up a polymer chain. After each reaction, you could filter and wash, and your peptide would stay attached to the insoluble support. You could wash it thoroughly and get rid of all the excess reagents, and you were ready to do the next step. You wouldn't have to take each peptide intermediate and purify it and crystallize it, which was the standard way to proceed. Then, when you had your sequence assembled, at the end, you could break the bond that was holding the peptide to the support. Of course, you had to have the right conditions under which this bond would break and the rest of the bonds would remain stable. Many ways have now been found to make the attachment and to break it."[3]

Merrifield had visible traces of plastic surgery on his face. He told us what happened and issued an important warning.

"When I was a teenager, I had an infection on my leg and I went to a dermatologist who was using a brand new X-ray machine. He cured the leg, but then he said, you have some acne on your face, why don't I cure that too? So he gave me X-ray treatment for that. It was fairly early in X-ray treatment, and the doses were very large. Then about fifteen years later, tumors began to show up, and they have kept growing, and I have two or three a year that have to be removed. Don't ever let your kids have X-ray treatment unless it's absolutely critical."[4]

---

[3] *Candid Science III*, 208, 210–11.
[4] Ibid., 217.

GEORGE A. OLAH

People with scientific backgrounds have a role to express themselves.

George A. Olah and Istvan Hargittai in 1996 at the University of Southern California.
(photo by M. Hargittai)

George A. Olah[1] (1927–) is a Hungarian-born American chemist who in 1944–1945 miraculously survived Nazi persecution in Budapest. He was among the many children saved by Pastor Gabor Sztehlo. In 1949, Olah received his doctorate at the Budapest Technical University and started his research career there. In 1957, following the suppression of the anti-Soviet Hungarian Revolution, he moved to Canada, and then to the United States. Since 1977, he has been at the University of Southern California. In 1994, he received the Nobel Prize in Chemistry "for his contribution to carbocation chemistry."[2] We recorded our conversation in 1996 at USC.

---

[1] I. Hargittai and M. Hargittai, *Candid Science I*, 270–83.
[2] "The Nobel Prize in Chemistry 1994," *Nobelprize.org*. Nobel Media AB 2013, http://www.nobelprize.org/nobel_prizes/chemistry/laureates/1994/.

Here we quote what Olah said about the controversy in organic chemistry that his research helped to solve and that gave his work great visibility.

"I wish I could tell you that this was an idea deus ex machina, or that this came as a revelation in my sleep. In reality, it was a long process. In science we need to have both concepts and facts. Science still is very much based on findings. It was a long thought process putting together many observations going back to the classical/nonclassical ion controversy, which was one of the major chemical controversies of our time, which allowed me to realize this rather fundamental new aspect of organic chemistry.

Was this controversy truly so important?

No, per se it was not. Frankly, whether the norbornyl cation has a bridged structure or an open equilibrating structure, this would have, per se, not affected in any way the future of chemistry. It would have, however, certainly affected the egos of the involved scientists. I came into it because around 1960 I discovered methods to generate positive organic ions, called now *carbocations*, as long-lived species, and we were able to take all kinds of spectra and establish their structure, including that of the norbornyl cation. In the course of this work I realized, however, that the problem has much wider implications. In the norbornyl ion the C–C single bond acted as an electron donor nucleophile. In this particular case this happens within the molecule, that is, intramolecularly. This delocalization, which had been originally suggested by Winstein, was indeed there and we were able to see it directly for the first time. Later came, what I thought was a logical idea. The question...I asked myself one day was, if this can happen within the molecule, why can't it happen between molecules? This led to the discovery of a wide range of electrophilic reactions of saturated hydrocarbons, that is, of C–H or C–C single bonds and the realization that carbon, under some conditions, can indeed bind five or even more neighboring groups.

When was this?

This was in the mid- and late sixties, and it was further generalized in the early seventies.

Were you a latecomer into the controversy?

I wasn't a latecomer; it was just that the controversy had started some years before I got interested in it. It started in the mid-fifties. I discovered what are now called *stable ions*, superacidic chemistry, in the late fifties, early sixties. So I was in the right position to enter and decide the issue. People who had been 'fighting' this controversy before had only indirect methods at their disposal, such as kinetics and stereochemistry. We, in contrast, could prepare stable, long-lived carbocations and study them directly by spectroscopy and other means.

In a nutshell, the controversy arose from differing attempts of explaining the same experimental facts. Winstein and others found that if they solvolyzed esters of norbornane (bicyclo-2.2.1-heptane), the exo and the endo isomers reacted with different rates. The exo isomer consistently reacted hundreds of times faster than the exo isomer. Winstein suggested that the explanation for this is that the exo reaction is accelerated. With great foresight, he indicated that the reason for this was neighboring C–C bond participation. In other words, solvolysis is accelerated because in a carbon-carbon bond, the electrons stored between the C_1 and C_7 atoms can interact with the developing carbocationic center. On the other hand, H. C. Brown, while looking at the same data (which incidentally were never in doubt, and this is significant) as Winstein, however, said that there is no acceleration or bridging in the intermediate ion. He suggested that the high exo/endo rate ratio is not due to the fact that the reaction of the exo is fast but, in fact, endo is slow. Because Brown was always much interested in steric interactions, he suggested that the reason for this is that the approach at the endo side of the molecule is hindered. The intermediate carbocation was considered by Brown as a pair of rapidly interconverting classical norbornyl cation, and not as Winstein's bridged nonclassical cation.

Here is an example where people were looking at the same set of data. They made very sure that the data were correct, but one explained the rate ratio difference of $k_{exo}/k_{endo} = 300$, saying that the exo rate is faster by so much, and the other said, no, the same result is due to the fact that the endo is slower due to steric reasons. It was the interpretation of the data which was argued, not the experimental facts. In this regard, I have a favorite quotation from a fellow Hungarian and Nobel laureate who won his Nobel Prize in medicine for his studies of the inner ear, George Bekesy. He wrote that what all scientists need is to have a few good enemies. When you do your work and write it up and you send it to your friends asking for their comments, they are generally busy people and can afford only a limited amount of time and effort to do this. But if you have a dedicated enemy, he will spend unlimited time, effort, and resources to try to prove that you are wrong. Bekesy ended up saying that his problem in life was that he lost many of his good enemies who became his friends. I don't want to comment on whether the situation is the same in chemistry or not, but I can assure you that many people have worked hard in this field to try to find experimental errors, and enormous effort was spent to establish and check the facts. I believe, in contrast to many, that this was no waste. One starts out to do research, and initially you may have no idea of the importance of your interest. I came to this problem because I had a method which enabled me to look directly at intermediate carbocations. People challenged me and told me. If you can look at these ions, tell us, in the norbornyl controversy, is Winstein or Brown right? I did it, which, per se, was perhaps rather simple. In the process, however, I found the general basis of the electrophilic reactivity of σ-bonds which can not only be intramolecular (as σ-participation in the norbornyl cation) but can involve σ-bonds in intermolecular (between molecules) reactions. This represented the key for electrophilic reaction for alkanes (including their C–H or C–C bonds). We eventually even challenged methane, the simplest hydrocarbon molecule, and we found that you can protonate (or otherwise attack by electrophiles) methane. The intermediate CH_5^+ is not only real but there is a broad chemistry of CH_5^+ and its homologues. Coming back to your original question, I was fortunate to do something quite significant in this field."[3]

---

[3] *Candid Science I*, 272–76.

LINUS PAULING

My arguments about the chemical structure... pertain to all molecules and crystals.

Linus Pauling in 1984 at Moscow State University. (photo by and courtesy of Larissa Zasurskaya, Moscow)

Linus Pauling[1] (1901–1994) was one of the greatest chemists of the twentieth century. He is the only person who has won two unshared Nobel Prizes; one in Chemistry in 1954, and the other in Peace in 1963 (for the year 1962). One of us (IH) had several encounters with Pauling toward the end of his life. Here, IH narrates: The exceptional Norwegian chemist, Otto Bastiansen, was one of my mentors. Bastiansen did his postdoctoral work with Linus Pauling around 1948 in Pasadena, and Pauling had a long-lasting impact on Bastiansen. Therefore, I could consider myself a scientific descendant of Linus Pauling. When in the fall of 1993, I was considering launching the magazine *The Chemical Intelligencer*, I solicited the opinions of a few famous chemists about such a project. Pauling was one of the most enthusiastic supporters, although

[1] I. Hargittai and M. Hargittai, *Candid Science I*, 2–7.

he regretted that he could not contribute to the magazine because he was so busy. It was not that he was so busy at the time; rather, he was terminally ill. His expression of regret made me come up with the idea of a brief interview, which eventually became the seed for our whole new interviews program. I sent a few questions to Pauling and he responded promptly.

We asked Pauling about the two recent discoveries of new materials, C_{60}—buckminsterfullerene—and quasicrystals. Buckminsterfullerene consists of sixty carbon atoms situated in the vertices of the truncated icosahedral shape, and its initial experimental observation in 1985 made quite a stir in the chemistry community.

"I am rather surprised that no one had predicted the stability of C_{60}. I might have done so, especially since I knew about the 60-atom structure with icosahedral symmetry, which occurs in intermetallic compounds. It seems to be difficult for people to formulate new ideas. An example is that from 1873 to 1914 nobody, knowing about the tetrahedral nature of the bonds of the carbon atom, predicted that diamond has the diamond structure."[2]

As is known, in 1996, for the first experimental observation of C_{60}, R.F. Curl, H. W. Kroto, and R. E. Smalley were awarded the Nobel Prize in Chemistry. Not only Pauling but also Curl, Kroto, and Smalley were unaware of the earlier predictions of a 60-member carbon cage of truncated icosahedral shape, made by Eiji Osawa in 1970 and by Elena Galpern in 1973. Osawa came up with his suggestion purely based on symmetry considerations. Galpern carried out rudimentary quantum chemical calculations from which she predicted the stability of such a structure. To Pauling it seemed very obvious that such a structure should exist.

The situation with quasicrystals was very different. Quasicrystals are structures intermediate between the classical crystal structures and the amorphous bodies. The former are regular and periodic; the latter are neither. Quasicrystals are regular but they are not periodic. According to the rules of classical crystallography they could not exist. In a simpler way, this is expressed by saying that fivefold symmetry is forbidden in crystal structures. This was a long-held dogma, and Dan Shechtman's experimental observation in 1982 destroyed it. Soon there were convincing theoretical models interpreting Shechtman's observations. In 2011, Shechtman received the Nobel Prize in Chemistry for his discovery. Pauling, to the end of his life, did not believe that the interpretation of Shechtman's observation was correct (he did not doubt the validity of the observation itself).

"As to the quasicrystals, you know that I contend that icosahedral quasicrystals are icosahedral twins of cubic crystals containing very large icosahedral complexes of atoms. It is not surprising that these crystals exist. The first one to be discovered was the MgZnAl compound reported by my associates and me in 1952. We did not observe quasicrystals of this compound, but they have been observed since then."[3]

---

[2] Ibid., 5.
[3] Ibid., 6.

In 1939, Pauling published his extremely influential book The Nature of the Chemical Bond *and in 1960 the third edition of the book appeared. Neither Pauling, while he was alive, nor anybody else since has taken up the formidable task of trying to update the third edition. Here is what in 1993 Pauling responded to the question whether the ever improving computational techniques would make his generalized observations in structural chemistry outdated.*

"I do not think that quantum mechanical calculations of molecular structure or crystal structure will ever make the sort of chemical arguments about structure presented in my book obsolete. The quantum mechanical calculations are made for one substance, and perhaps then for another somewhat similar substance. My arguments about the chemical structure are very general and pertain to all molecules and crystals."[4]

---

[4] Ibid.

JOHN C. POLANYI

Science is having a colossal effect on the world scene, and... we cannot responsibly opt out of the debate on world affairs.

John C. Polanyi in 1995 at the University of Toronto. (photo by I. Hargittai)

John C. Polanyi[1] (1929–) was born in Berlin, Germany, was four years old when he and his family became refugees from Nazi Germany and settled in Manchester, England. In 1940, his parents—his father was Michael Polanyi, the internationally renowned chemist turned philosopher—sent him to Canada to protect him from the German bombings. Upon his return to England he completed his high school studies and continued at Manchester University, where he earned his PhD degree. From 1956, he has been at the Department of Chemistry, University of Toronto, in Canada. In 1986, he shared the chemistry Nobel Prize with Dudley R. Herschbach and Yuan T. Lee "for their contributions concerning the dynamics of chemical elementary processes."[2]

---

[1] I. Hargittai and M. Hargittai, *Candid Science III*, 378–91.
[2] "The Nobel Prize in Chemistry 1986," *Nobelprize.org*. Nobel Media AB 2013, http://www.nobelprize.org/nobel_prizes/chemistry/laureates/1986/.

John Polanyi has been a public figure in Canada. We recorded our conversation in 1995 in his office at the University of Toronto.

On his engagement in politics.

"I've been engaged in various political debates, such as those sponsored by Pugwash, for over thirty-five years. My experience is that one can readily get access to senior politicians. The real question is, can one get them to listen? I would claim that if you make a sufficiently cogent argument they will have to listen. I think we scientists have some training in organizing our thoughts and in trying to persuade difficult audiences.

In fact, our colleagues constitute just such an audience. When you go to a scientific meeting with a new idea, you don't expect people to applaud. What they do is to tear it apart. That's their function. In science, we arrive at the truth through an adversarial dialogue. So we are used to having to make a case, and shouldn't be frightened to do so.

In the past, it is true, we were frightened to speak out on larger issues. We felt that we'd be trafficking in our reputation as scientists in order to get a hearing and that as a result we could bring science into disrepute, since people could say that we had abused our credentials.

As with most criticisms, there is something to this one. We do have to be careful about this. We have to explain what our expertise is.

But were we to take the opposite view that science is a sort of priesthood and that to keep it in high esteem we must keep it pure by ensuring that no scientist or group of scientists meddles in things outside their own discipline, then we would be involved in a different sort of irresponsibility.

The fact is that science is having a colossal effect on the world scene, and as a result we cannot responsibly opt out of the debate on world affairs. Earlier today, I had somebody in here who wanted to talk about the fiftieth anniversary of the dropping of the first atomic bomb in Japan. I was sixteen in 1945 when that bomb was dropped. Though it came at the end of a huge and terrible war, it was a transforming moment in other respects, too. My own thinking was deeply affected by it, as it should have been. It transformed, for example, the relations between nations. But it was only one of a whole series of technological changes, to which we scientists have contributed, that have changed the world. It would be irresponsible, therefore, for us not to be involved in the debate that follows.

Often, all that we have been able to contribute has been technological solutions. But even these can have tremendous impact. I have, for example, been involved in a lot of arms control discussions with Russian scientists. They may perhaps seem peripheral now, but at the time two colossal adversaries were threatening each other and the world. The danger of the arms race was very real. Nonetheless, the political community, our leaders, were saying that we can't do anything about it for a whole lot of technical reasons. It was necessary for the technical community to say quite specifically, 'yes, we can.' If we fail to stop testing of nuclear weapons, or if we fail to reduce the number of nuclear weapons, it's not because it's impossible to verify these things. We explained in some detail how we could do it. If then we *didn't* do it, it was because we didn't want to. By clearing the way on the technological level, one has an undoubted influence on the way history unfolds."[3]

---

[3] *Candid Science III*, 384–85.

JOHN A. POPLE

I was a curious child and started a research program when I was 12.

John A. Pople (on the right) and Walter Kohn in 2001 in Stockholm. (photo by I. Hargittai)

John A. Pople[1] (1925–2004) was born in Burnham-on-Sea, Somerset, England. In 1943, he won a scholarship to Trinity College, Cambridge University, where he earned degrees in mathematics and, in 1951, his PhD degree. He moved to the United States in 1964 and was affiliated with Carnegie Mellon University in Pittsburgh, Pennsylvania, until 1991. After that, he was at Northwestern University in Evanston, Illinois. Pople shared the 1998 Nobel Prize in Chemistry with Walter Kohn. Pople's award was "for his development of computational

---

[1] I. Hargittai and M. Hargittai, *Candid Science I*, 178–89.

methods in quantum chemistry."[2] We recorded our conversation three years earlier, in 1995, in his office at Northwestern.

We asked Pople to explain what computational chemistry is.

"Computational chemistry is the implementation of existing theory, in a practical sense, to studying particular chemical problems by means of computer programs. Some people draw a distinction between computational chemistry and the underlying theory. I really prefer not to, and think of computational chemistry as implementation of theory for the understanding of chemical problems. The theory has preexisted, and the computers have enabled it to be implemented much more broadly than was possible before.[3]

Should we treat computational chemistry as a separate discipline?

There are professors of theoretical chemistry. However, I don't think computational chemistry should be a separate discipline. It's a technique that all chemists should use. So it should be in the general curriculum; it should be taught but not... necessarily by a theoretical chemist. The programs should be considered a black box, just like a complicated spectrometer is, and chemists should learn how to use these programs, and to use them in a critical manner, to understand the limitations of what they get out, just like any other technique."[4]

About Pople's move to the United States.

"My early research career was in England. I started working in chemistry in 1948 under Lennard-Jones as a graduate student, and held appointments in England until 1964. I was on the Faculty of Mathematics in Cambridge and taught mathematics rather than chemistry. I got my degree in mathematics. That's traditional in Cambridge. People do mathematics and then do theoretical science. This goes back to Isaac Newton. Many of the best known theoretical physicists in Cambridge, people like Dirac, were professors of mathematics. I was a lecturer of mathematics. Then from 1958 to 1964, I worked for the National Physical Laboratory, which is the British equivalent of the National Bureau of Standards in the US. I moved to the Carnegie Mellon University in 1964, as Professor of Chemistry. Part of the motivation to move here was the better audience in America for computational science than in England. The very first time I visited the United States, in 1955, when I was thirty, I had never given a lecture in a chemistry department in England. I had given seminars within the theoretical chemistry group but never a general lecture. Then I came to this country and made a tour of several universities, including UCLA and Chicago. I found it quite remarkable to get a hundred people to come to listen to what I had to say. The US was the best place to develop theory at that time."[5]

---

[2] "The Nobel Prize in Chemistry 1998," *Nobelprize.org*. Nobel Media AB 2013, http://www.nobelprize.org/nobel_prizes/chemistry/laureates/1998/.
[3] *Candid Science I*, 180.
[4] Ibid., 184.
[5] Ibid., 185–86.

Pople came from a middle-class family, yet he studied at Trinity College of Cambridge University—a top venue considered to be rather exclusive.

"It's always been possible to get into Cambridge, or Oxford, from any background by a scholarship arrangement. I competed for a scholarship, sat for an examination at Trinity College, and became a Scholar of the College. This is a tradition that goes back a long way. People can become Scholars of the Cambridge Colleges, whatever their background, provided they make it through this examination. This gives you privileges when you get to the university. This way people of all sorts of background move into professional classes. It has been a way in which England is not really totally a class-ridden society. People who go to Oxford and Cambridge may go there because it's the thing to do or because their parents are upper class, but others come in as Scholars. Students are classified as Scholars because they had won a scholarship, and the rest of the students are called Commoners. The Scholars have some privileges. When you go to the College Chapel, for example, the Scholars sit in privileged places and the Commoners sit at the bottom."[6]

---

[6] Ibid., 188.

GEORGE PORTER

There is nothing more rewarding than linking two quite different subjects.

George Porter in 1997 at Imperial College in London. (photo by I. Hargittai)

Baron George Porter of Luddenham[1] (1920–2002) was born as George Porter in Stainforth, South Yorkshire, England. He was a physical chemist who studied the way chemical reactions happened. He was a much-decorated scientist, famous also for his popular lectures on television. In 1967, Porter received the Nobel Prize in Chemistry together with R. G. W. Norrish, his former mentor at Cambridge University "for their studies of extremely fast chemical reactions, effected by disturbing the equilibrium by means of very short pulses of energy."[2] Theirs was half of the prize—the other half went to Manfred Eigen. We recorded our conversation in 1997 in Porter's office at Imperial College in London.

---

[1] I. Hargittai and M. Hargittai, *Candid Science I*, pp. 476–87.
[2] "The Nobel Prize in Chemistry 1967," *Nobelprize.org*. Nobel Media AB 2013, http://www.nobelprize.org/nobel_prizes/chemistry/laureates/1967/.

Porter was most enthusiastic about science popularization. During a Christmas lecture at the Royal Institution, he had a beautifully laid table and after repeating three times the sentence, "I believe in Isaac Newton," he pulled out the tablecloth with everything intact on the table.

"Yes—illustrating the law of inertia. I would like to add something to that story. Those Christmas lectures were watched by the Royal Family, encouraged by the Duke of Edinburgh, at Buckingham Palace at Christmas. In the New Year, the Queen kindly invited me to lunch there, with her family and a dozen other people. Princess Ann was there; I've always liked Princess Ann; she is a bit of a tomboy. She was about seventeen at this time. She said, 'I liked your demonstrations, especially the one where you cleared the table.' We were walking toward lunch at the time, and as we entered the splendid drawing room with its large table set for twelve people and loaded with silver and gold plate, Princess Ann said, 'Go on, try it now!'

Like every other professor, I had to give an inaugural lecture when I became a professor at the University of Sheffield. I took a lot of trouble and I got together about twenty or thirty spectacular and interesting demonstrations. It went down extremely well, and someone told Sir Lawrence Bragg about it. He was then the director of the Royal Institution, and he invited me to lecture there to 350 school kids. Then I gave a discourse to the Royal Institution, quite a formal event, and a number of press and television people were there. It was about entropy and it was called, 'The Laws of Disorder.' I gave six one-hour lectures at a time when there was one television channel only, and they were on in prime time. That was how I got involved in presenting science to the public, which I greatly enjoyed.

Did you get through to the nonscientists?

I think I did. I remember coming into the London airport, very late one night, into a large, empty, hanger-like space. There were some other porters (those who carry luggage!) and one of these toughies shouted, 'What's new in thermodynamics?.' I did about 200 programs, and became quite well-known to the public in the early 1960s. However, instant fame on television doesn't last long.

This was before I went to the Royal Institution in 1966, where I stayed as Director for twenty years. I had a large research group and was very involved in the lectures for the public and for young people on the theories made famous by Davy and Faraday. Then, in 1985, I became President of the Royal Society. At a rather advanced age, I was invited to come here to Imperial College to be a professor. I have a nice new laboratory here. I am now thinking of taking 'early retirement,' at seventy-eight. It doesn't mean I'm giving it up, but I really can't expect Imperial College to support me forever. I am all for going on as long as I can, but there comes a point where you slow down a little bit. People of my age don't often do anything very bright or original themselves, although they can sometimes be useful when working with younger colleagues."[3]

---

[3] *Candid Science I*, 480–82.

VLADIMIR PRELOG

I am always strongly impressed by fateful consequences of often inconceivable sequences of events.

Vladimir Prelog in 1995 at the Swiss Federal Institute of Technology with a gold-plated model of the backbone of transfer ribonucleic acid molecule. The model was a gift from Alex Rich of the Massachusetts Institute of Technology, who had determined its crystal structure.
(photo by I. Hargittai)

Vladimir Prelog[1] (1906–1998) was an organic chemist. He was born in Sarajevo, Bosnia-Herzegovina. He attended high school in Osijek, Croatia, and was a student of the Prague Institute of Technology in Czechoslovakia, from which he graduated in 1928 and received his doctorate in 1929. Between 1935 and 1941, he taught at the University of Zagreb, Croatia. Then he moved to Zurich and remained affiliated with the Swiss Federal Institute of Technology for the rest of his life. Prelog and John W. Cornforth shared the 1975 Nobel Prize in Chemistry. Prelog's citation for his half read, "for his research into the

---

[1] I. Hargittai and M. Hargittai, *Candid Science I*, 138–47.

stereochemistry of organic molecules and reactions."[2] We recorded our conversation in 1995 in Prelog's office at the Institute.

Prelog remembered warmly his teachers and mentors. He studied directly under Ivan Kuria in Osijek, Rudolf Lukes in Prague, and Leopold Ruzicka in Zurich. Robert Robinson influenced him through his publications. Prelog had given the matter a lot of thought and finally decided that Ruzicka was the most important in his development.

"Sir Robert Robinson of Oxford was not my teacher in the usual meaning of the word, but I was very much influenced by his papers. Later, when I met him, I found that he was also an unusual person, not easy to deal with. In spite of that, I admired him. For many years, until about two weeks ago, his photograph decorated this room; but I replaced it with a portrait of Ruzicka because I realized that in fact he influenced me much more than Sir Robert did.

After my arrival in Zurich in December of the fateful year 1941, Ruzicka assigned to me the investigation of the lipophilic extract from 5000 kg of boar testes, which he had ordered from Wilson laboratories (next to the famous slaughter houses) in Chicago. He hoped it would be possible to isolate from it biologically interesting compounds. The results of my work were rather disappointing. The only exception was the isolation of small quantities of a compound with a musk-like scent. I determined its structure as 3-hydroxyandrosten-16. Ruzicka was interested in this result because the polycyclic steroid resembles structurally the monocyclic musk, civetone, which he had investigated at the beginning of his scientific career. Later, the compound became interesting for different other reasons. The same compound was isolated by other investigators from truffles. Since antiquity, sows are used to detect this delicacy under the ground, where it grows, by smell. Finally, it was also detected in the sweat of human males and is considered to be the mammalian male pheronome. It is used today in pig breeding as well as in perfumery.

Shortly after the war, Ruzicka changed considerably. For several years he neglected chemistry and spent most of his time collecting Dutch paintings. He had become a modest collector already during his short stay as a professor of chemistry at the University of Utrecht, but after the war he started collecting on a larger scale—thanks to chemistry. During the war CIBA Ltd. produced quantities of testosterone in the USA using Ruzicka`s method. A few million Swiss francs accumulated there as royalties, which the Americans did not want to transfer to Switzerland, as they feared that it would be occupied by the German army. After the war, when the money was transferred, a considerable part of it became subject to tax. To avoid this, Ruzicka donated it tax-free to a foundation which enabled him to create a collection of Dutch paintings exhibited at the Kunsthaus in Zurich. He concentrated his attention on buying paintings and spent most of his time visiting auctions in Switzerland and abroad, mainly in

---

[2] "The Nobel Prize in Chemistry 1975," *Nobelprize.org.* Nobel Media AB 2013, http://www.nobelprize.org/nobel_prizes/chemistry/laureates/1975/.

England. Because of all that Ruzicka was not any more the shining example I wanted to imitate.

There are many stories from this period of his life, of which I would like to tell what is perhaps the most fascinating one. The most valuable painting Ruzicka purchased in England was a portrait by Rubens of King Philip IV of Spain. In order to export such a painting from England, a permit from the National Gallery of London was necessary because this institution had the first option to buy the painting to keep it in England. After receiving such a permit, Ruzicka traveled to London to make final arrangements. During his stay there, he was invited by his friend Sir Ian Heilbron to a dinner, where he met the director of the National Gallery. They spoke about the permit, which had just been issued. The director mentioned that the painting was in a very bad condition and should be thoroughly cleaned. He also offered the service of his experts, an offer that Ruzicka gratefully accepted. They even agreed about the price. After cleaning it, the experts found that the painting was of remarkable beauty, and the National Gallery withdrew its export permit. Ruzicka was, of course, furious but he knew that, even if he could not export the painting, he could own it as long as it stayed in England. So he offered the portrait, then worth 750,000 Swiss francs, on loan to the Swiss Embassy in London. There it stayed until the diplomats arranged for its transfer to Switzerland, where it was exhibited as part of the Ruzicka Collection in Zurich. One day in 1985, an evidently frustrated and also insane young man from Munich went to the Kunsthaus, poured an inflammable liquid on the painting and burnt it to ashes. If Ruzicka had not been so successful in everything he undertook, this Rubens masterpiece could still be admired in England.

You may conclude from our conversation that I am always strongly impressed by fateful consequences of often inconceivable sequences of events."[3]

---

[3] *Candid Science I*, 144–47.

F. SHERWOOD ROWLAND

The people that I'm trying to provide answers for don't want to hear what the World Meteorological Organization has said.

F. Sherwood Rowland in 1996 at the University of California, Irvine. (photo by I. Hargittai)

F. Sherwood Rowland[1] (1927–2012) was born in Delaware, Ohio, and went to college locally at Ohio Wesleyan University. War service interrupted his studies, but he graduated in 1948. He went to graduate school at the University of Chicago where he had stellar professors in physics and chemistry. He received his PhD degree in 1952, started his career at Princeton University, continued at the University of Kansas, and from 1964 to the end of his life was at the University of California, Irvine.

[1] I. Hargittai and M. Hargittai, *Candid Science I*, 448–65.

Rowland had a distinguished research career, which was capped by his and his postdoctoral student Mario Molina's discovery about the role of chlorofluorocarbons (CFCs) in the depletion of the ozone layer. It was a long way from the discovery to its recognition and the Nobel Prize in Chemistry 1995, which Rowland and Molina shared with Paul Crutzen "for their work in atmospheric chemistry, particularly concerning the formation and decomposition of ozone."[2] Perhaps it was yet a greater triumph when in 1987 the Montreal Protocol on Substances That Deplete the Ozone Layer was signed. By the fall of 2009, all member states of the United Nations had ratified it. We recorded our conversation in Rowland's office in 1996 at the University of California at Irvine.

We asked Rowland whether he supposed that in addition to the scientific importance of their discovery, politics might have also played a role in their Nobel Prize.

"I think there is an element of scientific politics in it, but I don't think the awarding of the prize involved nonscientific politics. The question of whether or not this kind of work is fundamental chemistry, or whether it belongs in geosciences or meteorology, has always been there. To us it was always chemistry, but from 1974 until 1988, I did not get very many invitations to speak at chemistry departments. I'd get them from other university departments: toxicology, geology, physics. Part of this...is that there is a very strong feeling within most academic chemistry departments that this kind of work is not chemistry; rather, it is some sort of application only, because it's outside the laboratory. Another aspect is that chemistry has had a strong industrial component since the 1920s; and until 1988, we were directly in opposition to some of the major chemical companies, such as DuPont and Allied. In the 1974 time frame, two-thirds of the CFC production was in the form of propellants for aerosol sprays in the United States. About half of the global CFC use was then in the United States. The aerosol spray propellants were mostly chlorofluorocarbons. Some used hydrocarbons, but probably 80 percent of the aerosol industry was in the front line immediately, and they came out swinging. In the American Western movies, the good cowboys always wear white hats and the bad guys always wear black hats so the little kids know whom to cheer for. We had black hats, as black as you could get. Every month Molina and I would read their publication, *Aerosol Age*, to find out what else they were saying about us. One example that illustrates to me how far they would go was...an interview with a person who speculated that we were agents of the KGB, intent on disorganizing American industry."[3]

Even after the Nobel Prize, not everything seemed to settle down. This is how Rowland answered our question, "After all the difficulties, initial disbelief, and even accusations, today you are a Nobel laureate. Do you feel elated?"

"Obviously, any scientist is elated by recognition with a Nobel Prize. I wouldn't describe it as a goal, but rather the feeling that if you ever did anything really significant, such

---

[2] "The Nobel Prize in Chemistry 1995," *Nobelprize.org*. Nobel Media AB 2013, http://www.nobelprize.org/nobel_prizes/chemistry/laureates/1995/.
[3] *Candid Science I*, 462.

recognition was possible. One would expect that the judgment of the Swedes would persuade some people that what we did was a good thing, and in fact, that is the reaction of the overwhelming majority. However, in the US, it is just as controversial now, but essentially outside the scientific community. The backlash started in 1990. Legislation is being introduced, just as we speak, in several states permitting the use of CFCs. The immediate right-wing political reaction to the Nobel Prize was that the Swedes must really be in trouble for environmental politics to dominate things so much. The immediate assertion was, what a gross error the Swedes had made—there must be serious political problems in their country. A person from our Environmental Protection Agency asked me about a year ago, have you written anything up recently? I said, no, but the latest document of the World Meteorological Organization and NASA is out with all these details. And he said, the people that I'm trying to provide answers for don't want to hear that the World Meteorological Organization has said anything because the WMO is part of the United Nations. And, of course, they would be saying this because that is part of the takeover of the US government by the United Nations. So we are living in a very interesting period."[4]

[4] Ibid., 465.

FREDERICK SANGER

Most of our successes depended on recent successes of other workers.

Frederick Sanger in 1997 in Cambridge, UK. (photo by I. Hargittai)

Frederick Sanger[1] (1918–2013) was born in Redcomb, Gloucestershire, England. He was connected to Cambridge University for his entire professional life. His principal research achievements were the techniques he worked out for sequencing proteins and nucleic acids, and for each of these contributions he was awarded a Nobel Prize in Chemistry, the first in 1958 and the second in 1980. We recorded our conversation in 1997 in Cambridge.

It is hard to imagine how one embarks on projects that lead to such discoveries. Could it be that he had just set out these two great problems and then solved them?

"It was not quite like that. In fact, it all went by stages. I had got my PhD with Albert Neuberger on protein metabolism. In that work I learned a lot about protein chemistry.

---

[1] I. Hargittai (Ed. M. Hargittai), *Candid Science II: Conversations with Famous Biomedical Scientists* (London: Imperial College Press, 2000), 72–83.

In 2000 in Sanger's garden with Istvan Hargittai. (photo by M. Hargittai)

I started off, by luck, working on proteins. I happened to get a job in 1943 with Professor A. C. Chibnall, who was the new Professor of Biochemistry in Cambridge. He suggested to me that I should try to look at the end groups of insulin, that is, at the amino acids at the end of a polypeptide chain. Chibnall was interested in the number of amino acids in proteins. Nothing was known at that time on sequencing. People had tried to do it but had not made much progress.

I think the reason for choosing insulin was that it was a protein, probably the only one that you could buy in a pure form. Chibnall had done a lot of analysis on insulin. There was this interesting fact that it had a lot of free amino groups in it. He put me on this problem of trying to identify these amino groups. I was successful in developing a general method for looking at free amino groups. It was called the DNP method (DNP = dinitrophenyl). You put a colored reagent on the free amino group at the end of the chain and then you hydrolyzed the protein and identified the nature of the DNP-amino acid. The DNP was linked to the amino acid by a stable bond. The peptide bonds in the chain were broken down by acid. In this way you could identify the end groups. The main breakthrough that made this possible was due to previous work by A. J. P. Martin and R. L. M. Synge—the discovery of partition chromatography, which I applied to separating the DNP-amino acids. The work of Martin and Synge gave me the break. It was a very powerful fractionation technique. Previously crystallization and distillation were used for separating the amino acids.

So, you see, I didn't set out to solve the problem of sequencing. Rather, I just set out to determine the end group and worked out a general method to do it for proteins in general. I found there were two chains in insulin. One had phenylalanine at the end, and other had glycine at the end. One problem was that the DNP-glycine was rather unstable. When you did a complete hydrolysis, you didn't get a very good yield of the DNP-glycine, as it was broken down, so we had to cut down the time of hydrolysis. Then we found that we got a lot of other compounds produced. These turned out to be DNP-peptides. We looked at these and realized that we could get information about the sequence. With some work we could see two sequences about four or five residues long. Those were the first sequences determined in a protein. We had the two chains of insulin, and we could separate the two chains. Our next achievement was to determine the complete sequence of the phenylalanine chain, thirty amino acids long, by breaking it up into small fragments, fractionating them, and looking at their structure. We were again helped by the work of Martin and his colleagues in a new development of partition chromatography. That was paper chromatography, which enabled us to fractionate amino acids or peptides on a sheet of paper in two dimensions. Eventually, we were able to put the pieces together and determine the complete sequence. Martin and Synge were working at the time in Leeds, and Chibnall knew them well. Martin was the genius behind this. He discovered partition chromatography and went on to discover gas chromatography. He was a very inspiring person. I met him at meetings and he always had something new to talk about.

An important contributor to this work was a postdoctoral fellow, Hans Tuppy, who came from Austria after the war. He worked on the phenylalanine chain while I was working on the glycine chain. He was a very hard worker and finally got the sequence of the phenylalanine chain before we had finished the glycine chain. After that, he went back to Vienna, and has become more interested in administration.

When you received the Nobel Prize for the insulin work in 1958, you were forty years old. Then you set out another big task in research. This is not very common.

A lot of Nobel Prize–winners take on big administrative or teaching jobs or do something else. I'm not very good at teaching and I don't think I'd be very good in administration either. I don't think I would have enjoyed any of this the way I enjoy research. I had this opportunity, having gotten a Nobel Prize, to have a steady job and good facilities. It was easy to get students, particularly postgraduate students, who were well trained, and this helped very much with research. I was in a position to do more or less what I liked, and that was doing research.

Initially, I continued to study proteins. In 1958 nothing was really known about sequencing nucleic acids. It didn't seem to be an easy problem for two reasons. One was that they were so big, and the other was that they had only four components. A sequence with only four components would be very much more difficult to work out than a sequence with twenty components. For a while, I didn't see any hope of doing it, although I realized it was a very important problem. As DNA seemed quite impossible, I started thinking about RNA. The transfer RNAs were small, of about seventy or eighty nucleotides. It was around 1965 that I started working on RNA.

I used to go to the Gordon conferences in New England. There was always one on proteins and nucleic acids. I was bored by the nucleic acid talks and went largely for the proteins. Gradually, however, some of the nucleic acid talks rubbed off on me and I became interested in it.

The chief problem was...obtaining a small piece of RNA to sequence. The main progress was made by Bob Holley and his colleagues. They were able to isolate a pure transfer RNA and to sequence it, using the method which had been developed for proteins. So we were beaten to that one.

After that, we developed a two-dimensional method for fractionating small degradation products of RNA, which proved quite successful. It was a combination of ion-exchange chromatography and electrophoresis. We used this technique to study an RNA with 120 nucleotides in it, and this my PhD student George Brownlee managed to sequence. This was the largest system to that date, around 1965. We devised a few new techniques for sequencing the products, and this got us into the nucleic acid field."[2]

---

[2] *Candid Science II*, 76–80.

GLENN T. SEABORG

The role of chemistry in the biological sciences may determine our future.

Glenn T. Seaborg in 1994 in Anaheim with Istvan Hargittai.

Glenn T. Seaborg[1] (1912–1999) was born in Ishpeming, Michigan. He spent much of his professional life at the University of California, Berkeley, and the Lawrence Berkeley National Laboratory. His first job was personal assistant to Gilbert N. Lewis, whom he revered throughout of his life. Seaborg was co-recipient with Edwin M. McMillan of the 1951 Nobel Prize in Chemistry. Seaborg formulated the actinide concept of the heavy-element electronic structure, which correctly predicted the location of the actinide elements in the Periodic Table of the Elements. He participated in the discovery of many new elements and isotopes. Element 106 is named after him: Seaborgium, Sg. We recorded a conversation in 1995 in Anaheim, which was augmented in 1998, not long before Seaborg died. The excerpts below are from the 1998 supplement.

[1] I. Hargittai and M. Hargittai, *Candid Science III*, 2–17.

Seaborg's predictions for chemistry in the twenty-first century.

"Much progress will be made in the chemistry of life processes—in biochemistry, molecular biology, and related areas concerning the study of proteins, enzymes, nucleic acids, and other macromolecules. Chemical and biological investigations at the molecular and cellular levels, aided by enormously efficient computers, will elucidate the origin of life and perhaps lead to the artificial creation of life. Biochemical genetics will give us a great deal of control over the genetic code and, beneficently applied (which will pose a real challenge), should result in a reduction or elimination of genetic defects.

Immunochemistry, computer-aided molecular medicine, and chemotherapy should lead to the alleviation, treatment, cure, or prevention of our major ailments, including mental illness, and to a slowing of the aging process. We will understand the structure, mechanism, and functioning of nerve and brain, leading to control of our memory, through investigations in neurochemistry and the related fields of neuroanatomy, neurophysiology, statistical biology, and experimental psychology. Biochemical engineering should make available implantable (microcomputer-assisted) artificial hearts, kidneys, eyes (instruments to permit the blind to 'see'), ears (instruments to permit the deaf to 'hear'), and other bodily parts and organs.

Medicine underwent a revolution in the middle of this century, when it became possible to cure some infectious diseases with antibiotics. Now medicine is undergoing another revolution based on the applications of molecular biology—the mapping, cloning, and study of human genes to understand the body's normal functions at the molecular level, made possible by the unraveling of the role of DNA by James Watson and Francis Crick in the 1950s. A consequence of this will be gene therapy, one of the most exciting ramifications of molecular biology. The first successful gene therapy in a mammal was reported in 1984, when researchers injected a growth hormone gene into a fertilized mouse egg to overcome a genetic deficiency of growth hormone. The power of recombinant DNA technology lies in the opportunity it affords, for the first time, to produce virtually unlimited quantities of practically any protein and the promise it gives for the production of new drugs.

We are now engaged in a multiyear, multidisciplinary, technological undertaking to order and sequence the human genome, the complete set of instructions guiding the development of a living organism. The actual planned ordering and sequencing involves coordinated processing of some 3 billion bases from a reference human genome, and the consequent more complete understanding of human genes should lead to great advancement in the diagnosis and treatment of human diseases.

Another biological advance lies in monoclonal antibodies based on proteins produced by white blood cells in response to a foreign substance, such as a virus. Monoclonal antibodies are ideal for diagnosing infectious diseases. This technique, along with sophisticated computers, should make possible much more rapid diagnoses and, consequently, earlier selection of therapy.

Chemistry, properly supported, will help solve our energy, food, and mineral resource problems, even with our expanding worldwide population (hopefully at a diminished rate). Our endless supply of solar energy will be put to practical use through processes yet to be discovered—direct catalytic conversion to electricity, splitting of water to produce hydrogen fuel, or widespread bioconversion of vegetation and waste products. Chemists will increase the efficiency of extraction of new sources of minerals, and materials scientists will synthesize substitute materials from more abundant supplies.

Chemists will synthesize millions of new compounds tailored for a wide spectrum of practical uses. Nuclear chemists will be involved in the synthesis of additional chemical elements, hopefully in the region of the superheavy elements predicted to exist in the 'island of stability.'

These are but a few of the areas in which chemical progress will be made, and I have placed emphasis on the practical applications. The advances in theoretical chemistry will be tremendous in scope, and I shall not even attempt to make any predictions."[2]

Thoughts on the future.

"An important factor in the future, transcending the science of chemistry, will be the new public attitudes toward basic science and science in general—that is, the growing attitude toward ethical- and human-values considerations. The focus of this concern often is not on the question of whether the work is worth doing but instead on whether its potential harmful impact may outweigh any good it could do—that is, whether the research or project should be initiated at all. This attitude is affecting work on energy resources and technologies, biological research, aircraft development, and advances in the social sciences and education. This is going to have an increasing effect on the support and conduct of science, and I think most scientists are recognizing this.

As in the other cases of new influence, it is going to have its good and bad effects. Essentially, it is vital that science does serve the highest interest of society and contribute to the fulfillment of human values. And I believe that the science community for the most part is acting very responsibly, and responsively, in this direction. In many areas of research, such as genetic experimentation, atmospheric work, and the effects of chemicals on human health and the environment, it has taken the lead in placing human concerns above all.

But it should be realized that while there are certain values and ethical codes of a universal nature, there are also values that are more closely associated with the tastes, likes and dislikes, habits, and culturally induced beliefs of various individuals and groups attuned to certain so-called lifestyles. In a democratic society—and particularly one of growing advocacy and activism—there are bound to be many conflicts over these. And science and technology, with their increasing influence on life in general, certainly will be caught up in many of them. If this is the case, it may be essential that we find a way to establish some broad codes of conduct and values by which we can use science and technology to maximize human benefits within a framework of some type of consensus value scale. It seems to me that we must do this in order to avoid

[2] Ibid., 13–15.

being paralyzed by a kind of case-by-case value judgment of all that we do. This does not mean that technology assessments and risk/benefit studies of individual concepts should not be conducted. Nor does it mean that science should not maintain a most profound sense of responsibility toward safeguarding society from possible errors on its part or misapplications of its work. It does mean, however, that we must find a way to avoid having a 'tyranny of parochial interests' when it comes to the possibility of advancing the general good through scientific progress.

Perhaps I can summarize by suggesting that future directions of chemistry, and science and technology in general, may be influenced by two broad goals: more fully establishing the boundaries—physical, environmental and social—in which we can operate; and providing the knowledge capital that will allow us to operate within them. That knowledge capital—a product of basic research—upon which we have drawn so heavily in the recent past and which we must replenish with new ideas, might also allow us to compensate somewhat for declining physical capital and higher cost resources.

Finally, a few general thoughts: Our success in chemistry, and science in general, over the past century, and especially the last few decades, has brought us to a high level of material affluence, but this success also has fostered many new problems for the world. It also has given many people the notion that science should move us toward a utopian, problemless, riskless society. But this is a false notion. We live and always will live in a dynamic situation, amid problems whose solutions will breed other kinds of problems, and in a society where the leaps of progress will be proportionate to the risks taken. Even within the bounds of a 'steady-state society,' a 'no-growth society,' or any other scheme of population-resource-energy equilibrium we might achieve, there always will be change and creative growth that will challenge the human intellect. There always will be dangers, risks, and increasing responsibilities that will drive us toward a new level of excellence in all we do or try to achieve. This is the process of human evolution at work, a process that started with man's ascendancy and will continue for some time."[3]

[3] Ibid., 15–17.

NIKOLAI N. SEMENOV

One is... to uncover the internal relationships of the enormous number of elementary particles. This would let us into the details of the fundamental organization of matter. The other direction is the investigation of matter of higher organization,... the most organized of all, the living matter. [Semenov prognosticating in 1965 about the development of science.]

Nikolai N. Semenov (*center*) and his associates around 1930 at the Institute of Chemical Physics in Leningrad. On Semenov's right, Viktor Kondratiev; on Semenov's left, Yulii Khariton; and on Khariton's left, Aleksandr Shalnikov—all future internationally renowned members of the Soviet Academy of Sciences. (courtesy of Semenov's grandson, Alexey Semenov)

Nikolai N. Semenov[1] (1896–1986) was a physicist who was one of the founders of chemical physics, applying physics to the understanding of chemistry. Relying on the experimental observations of his associates, he discovered the branched chemical chain reactions. This was a few years before a most

---
[1] I. Hargittai and M. Hargittai, *Candid Science I*, 466–75.

important branched nuclear chain reaction was discovered, serving as the starting point of the atom bomb. Semenov studied in St. Petersburg/Petrograd/Leningrad (different names of the same city; today, again, St. Petersburg). Eventually, he built up a scientific research empire in the Soviet Union. In 1956, he was awarded the Nobel Prize in Chemistry for his discovery of the branched chemical chain reactions. We recorded our conversation in 1965 in Budapest.

We asked him about the origin of his interest in science and about the building up of scientific research in the fledgling Soviet State in the 1920s.

"I have to refresh very old memories to answer this question. I was still in school, contacted typhus, and could not attend school for quite a while. Since I didn't want to be left behind, I read a lot of books. My favorites were all chemistry books. When I got well I got to the nearest drug store and bought all the available chemicals there and started experimenting with them. For me, it was the greatest puzzle that sodium, this flammable and malleable metal, and chlorine, this extremely reactive gas, formed the innocent table salt. To check this out, I bought a piece of sodium, burned it in chlorine gas, and recrystallized the precipitate. It was a white powder which I poured over a big slice of bread and it was table salt indeed, the best kind. Although I was not aware of it, I hit right on one of the fundamental and at that time unsolved problems of science. It became possible to solve only many years later using the electron theory and the atomic arrangements in molecules.

My interest in chemistry kept deepening until I read in a book that the future of chemistry was in physics and that for a good chemist it was mandatory to know physics very well. Since my goal was to become a good chemist indeed, I signed up for the faculty of mathematics and physics of the University. I was enormously impressed there by Academician [Abram] Ioffe and started working for him in my sophomore year. Our main interest was the electron-molecule collisions. We were very excited by Niels Bohr's new theory, which explained so many things.

When I graduated from the University, I immediately became Ioffe's assistant. At the same time there were tremendous changes in our country. There was a revolution, civil war was on its way, and so was foreign intervention. Many thought at that time that under the circumstances scientific research would be suspended. Due to Lenin's foresight, however, things took a different turn for science. He convened the scientists scattered all over the country, created conditions for independent work, and secured the necessary funds for international monographs and journals.

This was a heroic era indeed. It was moving and uplifting to see the thirst for science of our impoverished, tortured, and liberated people. We received letters from the most remote corners of the country. If somebody read or created something of interest, he let us know at once. The scientific institutes themselves turned often to the public for recruiting new coworkers. Although we could hardly pay them, people were joining us en mass. Let me just mention one example, [Nikolai] Chirkov, who is now a professor, abandoned animal husbandry for science, and he was not alone. Everybody was willing and ready, even if living merely on bread and water, to participate in creating the new science. It is also true, though, that the young researchers were given greater independence than is customary today. But we were all very young. Even Ioffe, the oldest among us, was not yet forty.

I was put in charge of an independent laboratory in 1920, and was already directing sixty coworkers by the early thirties. Progress was breathtaking and new institutes were mushrooming. Among them Pavlov's physiological institute, Ioffe's physical institute, institutes for radio-physics, radioactivity, and others. The technical physics institute was very strong, and I was in charge of its Department of Physical Chemistry. The reactions abroad were interesting. For quite a while, they could not appreciate what was going on in our country. Only when the spectacular results appeared—the atomic bomb, the atomic power station, the first Sputnik—only then did we finally receive their recognition.

However, it was a long, very long time before we reached that destination. In the 1920s, England and Germany were the great scientific powers. Our Soviet science was, to some extent, provincial, in spite of the many excellent scientists. American science was in a similar situation at that time. Today Soviet and American science are on the top. Without belittling the American achievements, their development was, no doubt, greatly accelerated by the tremendous import of outstanding European scientists. We, on the other hand, had to do everything ourselves; but, I think, we have solved our task with great success."[2]

---

[2] Ibid., 468–69.

FRANK H. WESTHEIMER

That would probably be James Bryant Conant [the strongest influence on his career].

Jeanne and Frank H. Westheimer in 1995 at Owl's Head on Squam Lake, New Hampshire. (photo by I. Hargittai)

Frank H. Westheimer[1] (1912–2007) was a pioneer in one of the branches of computational chemistry that subsequently became very popular: molecular mechanics. He was a physical chemist, then a physical-organic chemist, finally, a biochemist. He started at the University of Chicago, and when World War II came, he participated in the Metallurgical Project as part of the defense effort. Later, he had a most distinguished career at Harvard University. In the 1960s, he chaired a committee of the US National Academy of Sciences, subsequently

---

[1] I. Hargittai and M. Hargittai, *Candid Science I*, 38–53.

known as the Westheimer Committee, to determine the goals of chemistry in the service of the nation. We recorded our conversation in 1995 in the Westheimers' summer home at Owl's Head on Squam Lake, New Hampshire.

In our conversation, the most remarkable segment was when the question came up about who was the single strongest influence on his career.

"Aside from my father, that would probably be James Bryant Conant. I went to graduate school at Harvard in order to work with him. But Conant became President of Harvard in 1933, when I had been with him for less than a year, so I transferred to Elmer Peter Kohler, who assigned me a research problem that turned out quite differently than he had expected. Kohler came by once a week, and asked me, regularly, 'What did you do last week? What are these set-ups on your lab bench? What are you planning to do next week?,' and I'd tell him. Then he would literally snort, turn, and walk out. I never really knew what that was all about. But whatever it was, he didn't instruct me what to do next; he let me work out my own salvation. I owe my independence largely to Kohler because he didn't press me to do what he thought would be best.

Near the end of my PhD research, I applied for a National Research Council Postdoctoral Fellowship. There weren't many postdoctoral fellowships in those days, no government financing of research, and not much industrial support either. But the National Research Council offered about a dozen postgraduate fellowships in chemistry, and I won one of them to work with Louis Hammett, one of my heroes in physical organic chemistry, in his laboratory at Columbia.

About that time, Conant called me into his office. He said that he knew I was getting my doctorate and was interested in my career. What was I going to do? I told him I'd won this fellowship and explained with great pride the problem I'd submitted and was going to work on. Conant had the habit of putting the tips of his fingers together and rocking back and forth while he thought and he put the tips of his fingers together and rocked back and forth, and then he said, 'Well, if you are successful with that project, it will be a footnote to a footnote in the history of chemistry.' As I walked out of his office, I realized what he had told me.

Really, it was two things. One was, of course, that my project wasn't very important. The other was—and it may have been pretty stupid that I had never thought this until that moment—that I was supposed to do important things. Chemistry was a lot of fun; it was great entertainment, and I was going to be paid—or at least so I hoped—for entertaining myself with it. Yet Conant had essentially told me that I was expected to do things that were scientifically important. The interview with Conant provided a vital kick in the pants for me. It changed the way I thought about my future.

At Columbia, I did the project that I had proposed, and it worked out beautifully. But it was exactly what Conant had said it was: a footnote to a footnote in the history of chemistry.

Then I set my sights higher—much too high, as a matter of fact. As a physical-organic chemist, I was concerned with general acid-general base catalysis and had decided that enzyme catalysis was probably caused by simultaneous general acid-general base catalysis. I was going to demonstrate this in my next piece of research. Amino acids, with their combination of acid and base in the same molecule, should prove especially active catalysts themselves. So I tried their catalysis of the mutarotation of glucose, but it turned out that there was nothing special about them. The project was obviously enormously ambitious, and although I was fundamentally correct about enzymes, demonstrating it was much too big a project for me at the time. That attempt came to nothing, but at least Conant wouldn't have been able to object that the attempt was directed at a footnote to a footnote. .

Eventually, I settled down to things that were more important than footnotes to footnotes, but not as grandiose as the youthful project that I just mentioned. I never discussed my research with Conant again, but I did restrict myself to things that he might have approved of.

Many years later, after I was a professor at Harvard, and after Conant had retired from his many careers, I was working in my office one Saturday when someone knocked on my door. I opened it, and there he stood. He looked at me and said, 'Do you remember me?' Needless to say, I did."[2]

Westheimer was so moved that following the sentence, "Do you remember me?" we had to end the conversation. The last sentence was added later, in our correspondence. I (IH) am often asked about the most remarkable moments of my conversations with famous scientists. This was certainly one of them.

[2] Ibid., 51–53.

ADA YONATH

The vision to understand ribosome was an irresistible intellectual stimulus.

Ada Yonath in 2004 in Budapest. (photo by M. Hargittai)

Ada Yonath,[1] née Livshitz, (1939–), was born in Jerusalem and received her PhD degree in structural biology from the Weizmann Institute in 1968. She established the first protein crystallography laboratory in Israel and spent extended periods in international laboratories, especially in Germany. In 2009, she was awarded the Nobel Prize in Chemistry together with Venkatraman Ramakrishnan and Thomas A. Steitz "for studies of the structure and function of the ribosome."[2] We recorded our conversation in 2004 in Budapest.

Yonath talked about her road to science.

"In Jerusalem, mathematics and physics were open to almost everybody who had good grades from high school, but they were very restrictive in chemistry because

---

[1] I. Hargittai and M. Hargittai, *Candid Science VI*, 388–401.
[2] "The Nobel Prize in Chemistry 2009," *Nobelprize.org*. Nobel Media AB 2013, http://www.nobelprize.org/nobel_prizes/chemistry/laureates/2009/.

chemists needed lab space.... Nonetheless, I applied for chemistry and thought that I would move to something else if I did not like it, or if I would not be admitted, but I was. This made me very happy. During the first two years we had plenty of organic chemistry, inorganic chemistry, and physical chemistry, but hardly any biochemistry. By then I had decided that my interest was in biochemistry and biophysics. I always reached my goals even though not always through the most straightforward ways. I ended up with a master's degree in Biophysics and continued for a PhD at the Weizmann Institute in Chemistry. I was considering doing protein crystallography, but it was not yet well developed. This was in the mid-1960s. Prof. W. Traub suggested to me to do fiber crystallography, and I worked on the structure of collagen.

After my doctorate, I went to Pittsburgh for a postdoctoral year at the Mellon Institute, and did muscle research. However, protein crystallography was still in my mind as my principal interest. So, finally, I started doing protein crystallography in the group of F. Albert Cotton at the Massachusetts Institute of Technology in Cambridge. This meant a major turn in my professional career. It helped also that I had extremely good relations with William Lipscomb at Harvard University. I had contacted him during my collagen work when I was considering coming to work with him, but at that time it did not happen. During my year at MIT, Lipscomb gave a course there, which I attended.

Following the two years in America, I returned to Israel, and started my own group in protein crystallography. I was alone in the entire country. I had an instrument and some limited lab space, and it took almost half a decade before things started working. During this period we did publish papers on simple structures.... I had to learn a lot of the procedures from books and at meetings.

From the beginning of the 1970s, protein crystallography in Israel was slowly moving to the frontier of science. It was also at that time that I developed a collaboration with Professor Michel Ravel at the Weizmann Institute. He had a technique to prepare what he called large amounts of the initiation factors, which initiate ribosome function in protein formation. The work involved an intensive collaboration with the late Professor Paul Sigler, who came to Israel for a year and a half. We had good interactions with the Chicago group; later I had spent a sabbatical year there. We were trying to grow the necessary crystals, but failed. During this year, at a meeting in Canada, Professor H. G. Wittmann of the Max Planck Institute in Berlin talked about the ribosome and reported the sequence of the initiation factors. We talked and he was interested in collaboration, and eventually I went to Berlin.

It was just a few months before I was supposed to go to Berlin with one of my students that I was riding my bicycle to the beach—it was in February, and the weather can be beautiful in February in Israel—and I fell down in the middle of the street. I had a brain concussion. I was brought to the hospital. My brain concussion was more or less over within two weeks, but there were side effects, which prevented me from flying. Also, I needed an operation. After everything was done, I went to Berlin about five months after the original plan.

When I arrived in Berlin, in November 1979, the initiation factors were almost ready, and in my 'free time' I discovered that they had very active pure ribosomes in huge amounts from several bacteria and I suggested using them for crystallization. The people were very supportive. I knew that many prominent scientists had failed before in attempts to crystallize ribosomes, and if I would fail too, I would be joining a distinguished group of luminaries, including Francis Crick, Jim Watson, Aaron Klug, Alex Rich, and others. I knew though that this was my big chance. I went about the project very carefully, since I assumed that the difficulties with ribosome crystallization stem form their heterogeneity as well as their tendency to deteriorate. First of all, I went back to the old literature and studied everything there was written about the ribosome, especially techniques developed for maintaining their integrity for relatively long periods, required for crystallization. I took advantage of procedures developed in the sixties by A. Zamir and D. Elson. I spent only two months in Berlin, but after I had returned to Israel, they kept sending me almost every week pictures taken by a light microscope (neither fax nor internet were available at these days). In three or four months, we had micro-crystals, which were much too small to be studied as single crystals, but gave a promising weak powder pattern. Then it took about four years to get the first diffraction patterns that didn't look like garbage. Our paper about the micro-crystals came out in 1980, and for the past quarter century I have been involved in elucidating the structures of ribosome.[3]

And a brief introduction to what the ribosome is.

"The ribosome is an organelle within the cell, which appears in thousands of copies in each living cell. It is a huge and dynamic assembly of proteins and RNA. This is where the genetic code is being translated from the nucleic acids to the proteins. In other words, it is the site for the synthesis of proteins resulting from translation of messenger RNA (mRNA). Transfer RNA (tRNA) that decodes the genetic code carries to the ribosome the cognate amino acids, to be incorporated into the growing chain. The biosynthesis of proteins happens with fantastic speed in the ribosome. When a chemist wants to create a peptide bond, it may take days and high-temperature and other extreme experimental conditions, whereas the ribosome can do this fast, within microseconds, and under mild conditions within the living cell. Also, scientists often make mistakes; the ribosome hardly ever makes mistakes.

In principle, any ribosome can read any genetic code. A ribosome in the human body can translate the genetic code of bacteria and vice versa. The ribosome is a factory for making proteins and can follow any genetic instructions. However, the ribosome of a higher organism—mammals and eukaryotes—is more complex than the ribosome of bacteria. The higher complexity is a consequence of additional tasks concerning regulations and selectivity, and it has to do with more interactions with the cell. The differences between bacterial and mammalian ribosomes are subtle. Even the active sites or areas near the active sites contain some differences. This is why ribosomal antibiotics can work. The antibiotics should impact the pathogenic bacteria only and not the patient, not even cause side-effects. Sometimes replacement of one single nucleotide can make the difference in the effects on bacterial and mammalian, that is, human ribosomes."[4]

---

[3] *Candid Science VI*, 391–93.
[4] Ibid.,394.

RICHARD N. ZARE

I like to tackle problems that matter.

Richard N. Zare in 1999 at Stanford University. (photo by I. Hargittai)

Richard N. Zare[1] (1939-) received his degrees from Harvard University. He started his career at the University of Colorado, then, he was at the Massachusetts Institute of Technology, finally, he moved to Columbia University, before settling down at Stanford University in 1977. He is a versatile chemist, very creative in the broad areas of analytical chemistry, and knows a great deal about its foundations in theoretical physics. He has been a much decorated and visible personality, one of the best known among chemists in the United States. We recorded our conversation in 1999 at Stanford University.

[1] I. Hargittai and M. Hargittai, *Candid Science III*, 448–59.

Richard N. Zare (left) and David W. Chandler in the laboratory. (courtesy of R. N. Zare)

Zare analyzed samples from a Martian meteorite found in 1984 on Antarctica. The meteorite hit the Earth some thirteen thousand years ago and stayed intact under frozen conditions. The question was whether there was any evidence in this meteorite pointing to the possibility of life on Mars.

"What a painful episode in my life, one not totally of my own planning. My research group was working with Kathy Thomas-Keprta of NASA Johnson Space Center on interplanetary dust particles and cluster particles. She asked us as a favor to look at two rocks that were given the Walt Disney character code names of Mickey and Minnie to keep track of them and tell them apart. My graduate student Simon Clemett and I did not know what they were, but how could you possibly refuse such a request from a collaborator and friend. After all, looking at particles that you could only see under a microscope was so much more of a challenge, so the mass-spectrometric analysis of these samples should pose little problem. Simon found that the samples contained polycyclic aromatic hydrocarbons (PAHs), a class of organic molecules we had detected previously in interplanetary dust particles and in many different meteorites. What was so peculiar and special about these PAHs was that they were concentrated in the rims that enclosed carbonate globules in this basaltic, volcanic rock. When I told Kathy, she was quite excited and told us that Goofy would be sent to us in the mail at once and that we should stop everything and devote ourselves to looking at Goofy. I said I was doing nothing of the sort until I knew more about what we were actually working on. It was then that she told me that David McKay, Everett Gibson, herself, and others at NASA Johnson Space Center were secretly studying these samples from a Martian meteorite. Secrecy was necessary, she said, because this project had not been authorized by NASA and because of what they thought they might be finding. Simon and I were told that in the iron-rich rims that surrounded the carbonate globules, they were observing magnetite grains that looked like they came from biological activity and transmission electron microscope images that resembled fossilized bacteria-shaped objects. These findings, all located together, seemed to point to the possibility that some primitive form of life had been inside this meteorite. This meteorite had fallen to Earth in Antarctica about 13,000 years ago and had only been classified quite recently, largely based on isotope ratios, as one of the few, rare meteorites that came from Mars. What a wonderful sample we had. It would be the closest I imagine that I would come to Mars.

Our contribution to this project was to find and characterize these PAHs. In themselves, they could be biomarkers of fossilized organic matter of long ago, or they could have come from some nonliving source, such as metal-catalyzed reactions of organics on hot surfaces. If we could establish these PAHs as indigenous to the meteorite, then they would be the first observation of organic molecules found from Mars.

The rest is history, as they say. We wrote a paper titled 'Search for Past Life on Mars: Possible Relic Biogenic Activity in Martian Meteorite ALH84001.' This paper was ultimately published in *Science* after a lengthy and useful peer-review process, but not before the story was leaked to the news media and NASA decided to hold a press conference in Washington, D.C. The reaction of many of my colleagues was to the press conference and to the hype surrounding it as opposed to the actual paper, which was to me quite disappointing. And the appetite for the press on this topic seemed insatiable. I refer to this period in my life as 'Mars madness.' It was exciting but also often unpleasant and stressful.

It was my intention in our *Science* paper to put forward a hypothesis—a hypothesis deserving serious examination, not a matter to be debated in the press but to be thoughtfully investigated. I believe that this hypothesis has not been refuted, but neither has it been confirmed. If anything, I think that the body of considered expert opinion is suggesting that the evidence we put forward is inconclusive, and many different parts of it may have different explanations. I have been gratified, though, that this work did play an important role in causing us to rethink the purpose of space exploration. Prior to this publication, speculation about life forms outside of Earth and how such could come about was almost exclusively the realm of science fiction. Today it has become an issue that can be scientifically investigated and tested. I am proud to have played some small part in this transformation. Certainly, these issues involve chemistry among other disciplines. Progress will depend in no small part on the creation and perfection of new instruments that will allow us to explore with ever-increasing power the world of the very minute."[2]

---

[2] Ibid., 458–59.

In 2012, we asked Richard Zare for an update[3].

"It has always been a puzzle to me how reduced carbon, that is, carbon making bonds with hydrogen, such as in polycyclic aromatic hydrocarbons, could find itself in a basaltic Martian rock. I had thought that at such high temperatures oxidized carbon, that is, carbon making bonds with oxygen, such as in carbonates, would be found instead. Recently, Steele and coworkers (including work from my laboratory) examined 11 Martian meteorites, spanning about 4.2 billion years of Martian history. Of these, 10 showed reduced carbon in association with small oxide grains included within high-temperature minerals. It seems that Martian magmas favored precipitation of reduced carbon species during crystallization. These findings do not rule out possible life in the past on Mars, but they further weaken the hypothesis that was put forward earlier concerning how to interpret the findings we made for the Martian meteorite ALH 84001."[4]

---

[3] E-mail communication from Richard N. Zare, July 18, 2012.
[4] See A. Steele et al., "A Reduced Organic Carbon Component in Martian Basalts." *Science* **337** (2012):212–15.

AHMED H. ZEWAIL

If somebody teaches you as little as just a letter, you owe it to him to become his slave.

Ahmed H. Zewail in 1997 at Caltech. (photo by I. Hargittai)

Ahmed H. Zewail[1] (1946–) was born in Egypt and received his first degree in Alexandria. He came to the United States to do his doctoral studies and received his PhD degree at the University of Pennsylvania under the supervision of Robin Hochstrasser. In 1999 he was awarded the Nobel Prize in Chemistry "for his studies of the transition states of chemical reactions using femtosecond spectroscopy."[2] One of his titles at the California Institute of Technology is that he holds a Linus Pauling professorship. We recorded our conversation in 1997 at the California Institute of Technology.

---

[1] I. Hargittai and M. Hargittai, *Candid Science I*, 488–507.
[2] "The Nobel Prize in Chemistry 1999," *Nobelprize.org*. Nobel Media AB 2013, http://www.nobelprize.org/nobel_prizes/chemistry/laureates/1999/.

Zewail described his education in Egypt.

"I received a very traditional education. It was important to get an education but it was irrelevant what direction I would choose later. There was no pressure from my family, for example, to get into business and to make money. We have a proverb that I heard all the time: 'A degree in your hand is like the power of wealth in your pocket.' It protects you, makes you a better human being, and makes you culturally rich. This was the tradition. Achieving good grades and getting into a good university was the driving force among my classmates. The hope was to become somebody some day. There was nothing fancy in our school, but there was serious emphasis on lots of homework and discipline.

I don't know what triggered my interest in becoming a scientist. I went to school and I was getting A's in science. During the precollege years I sat up some science experiments at home in my room. I had some test tubes, and would heat some substances, such as wood, in them. I could have created an explosion, but I didn't; I was lucky. I was intrigued by these experiments, especially when observing the substance changing from solid to a burning gas. My mother didn't mind my experiments; I think she thought that it was part of my studies.

Learning was everything. When I went to visit the University of Alexandria for the first time and I saw its beautiful, ornate buildings, the shrine of knowledge and learning, I had tears in my eyes. I was emotional about the 'house of learning' and by then I knew that one day I was going to become a scientist.

When I entered the University of Alexandria, I started with the sciences. The first year I could take four subjects, chemistry, physics, mathematics, and one elective, and I chose geology. The second year you could drop one of these four. After the second year, the top few students were allowed to specialize in one subject. There were seven such 'special students' in our class of about 500, and I was one of the seven. I chose 'special chemistry.' I graduated with distinction and first-class honor. As a special student, I had privileges, such as taking classes in the professors' offices, borrowing books from them, and they knew us by name. I got a fellowship, which was like a salary. When I graduated there was a position waiting for me since I was the so-called first student of the University. The idea was that I would go away somewhere, get a doctorate, and return to my position at the University of Alexandria. For my doctoral studies I went to the University of Pennsylvania. My research supervisor was Professor Robin Hochstrasser, who introduced me to state-of-the-art scientific research. I had a productive four and a half years there with a dozen or so papers, but as important, I was exposed to a large number of interesting research problems."[3]

About his experience of his start in the United States.

"When I came to the United States I was seeking to receive a better PhD education. The people at the University of Pennsylvania had not seen too many Egyptians before in this situation. It was different from someone coming from Britain, Germany, or

---

[3] *Candid Science I*, 500–501.

France. They didn't know my background and my culture. I was twenty-two, and for me it was also a culture shock. I grew up in the Middle Eastern culture. Friends are very important. You can borrow money from friends without writing it on paper. You can visit without calling. If you are having a crisis, your friends will spend hours with you and talk to you. When a good friend left Alexandria to Cairo for a week, we would go to the train station, kiss the guy, and might even have tears in our eyes.

The culture in the United States looked different to me, and work was the number one priority. There was an episode early after my arrival in 1969. It was snowing and I was wearing light shoes and I fell down. Every car passed by me. It was not that they were bad people, it was just that people minded their own business. In Cairo, traffic would have stopped, they would bring a chair in the middle of the street, and somebody would rush to me with tea with mint.

Speaking of the culture, I now laugh at some incidents in the beginning. When good friends are together in Egypt, we sometimes use an Arabic saying, 'I kill you' in verbatim translation. It's just teasing, of course. My English was very poor at the beginning, and often I would just translate words and expressions. This is how I told one of my fellow graduate students, 'I kill you.' I immediately noticed that something terrible had happened. In his mind, I just came from the Middle East, and he was sure I meant literally what I had said.

Another difference was the student/professor relationship. In the culture I came from, the professor is respected like a prophet of knowledge. There is a saying that 'If somebody teaches you as little as just a letter, you owe it to him to become his slave.' I arrived at the University of Pennsylvania, and I considered my professor to be a prophet of science. I found out that he liked coffee so I bought a coffee machine, got the coffee, made the coffee, and brought it for him in his office. All the other graduate students thought I was really weird. They were sure I wanted something special from our professor. When I arrived in the US, I gave the professor a gift that my parents had selected and wrapped for him, and I saw a similar reaction. By the way, I now like it when my students do these things if they wish!

I ignored the cultural difficulties and spent all my energy on learning. I only wanted to get the best education I could get. For four years in Philadelphia, I did not learn much about the city, but just knew my way around. I was, and still am, very excited about science. It was a paradise of science. In Alexandria, we had difficulties in finding journals, chemicals, equipments, etc.; here I could have anything. In Philadelphia, I made friends; some remained for over twenty-five years."[4]

[4] Ibid., 504–6.

SECTION 3
Biomedical Scientists

WERNER ARBER

Some people are against genetic engineering because they do not want to eat genes in their food.

Werner Arber in 2005 in Lindau, Germany. (photo by I. Hargittai)

Werner Arber[1] (1929–) was born in Gränichen in the Canton Aargau, Switzerland. He graduated from the Swiss Federal Institute of Technology (ETH) Zurich in nuclear physics in 1953. He received his PhD degree in biology from the University of Geneva in 1958. Following a few years in Geneva and a couple of postdoctoral stays in the United States, he joined the faculty of the University of Basel in 1971 and has been there ever since. He shared the 1978 Nobel Prize in Physiology or Medicine with Daniel Nathans and Hamilton O. Smith "for the discovery of restriction enzymes and their application to problems of molecular genetics."[2] We recorded our conversation in 2005 in Lindau, Germany, during the annual gathering of Nobel laureates.

---

[1] Istvan Hargittai and Magdolna Hargittai, *Candid Science VI: More Conversations with Famous Scientists* (London, Imperial College Press, 2006), 152–63.
[2] "The Nobel Prize in Physiology or Medicine 1978," *Nobelprize.org*. Nobel Media AB 2013, http://www.nobelprize.org/nobel_prizes/medicine/laureates/1978/.

Arber has long been known for his concern about sharing scientific knowledge with the public, including his views on the controversy about genetically modified food. We pointed out that there has been a lot of misunderstanding about this.

"That's true. I am one of those who from the very early times said that one should care about this problem and make some risk assessment, and devise even political guidance for meaningful and useful applications. I'm not saying that there is no problem whatsoever, but I think we should face these problems on a solid scientific and ethical basis. Some of the biologists who had been educated in the classical way, without understanding the molecular processes that go on in living organisms,...merely make claims about the big danger of genetic engineering. They say it is a big danger because it is not natural. This view is sometimes taken as the expert opinion by politicians in search for approval by the general public. Then comes the influence of the religious belief in the formation of people's opinion, and this is why religions could not just be ignored. If you believe that God created the world and that God is in love with the world, which is his creation, then, of course, you assume that the world was made the best that can be. So it should be an optimized situation. Therefore, if genetic engineering changes something in the genome, the result can only be worse. If God made the optimum of the human beings as well as the plants that human beings eat, and so on, then you don't have the right to change something because you would counteract the will of God.

What is the way out of this situation?

The only way out is to have a debate. We should try to reconcile different views rather than to enhance confrontation. In addition, some people are against genetic engineering because they do not want to eat genes in their food. They don't believe that they eat genes in their daily food, and that's the way they understand nature.

We would have thought that the Asilomar meetings and what followed demonstrated that scientists recognize their responsibilities so that should have at least alleviated some of the fears of genetic engineering.

I was at the Asilomar Conference in February 1975, and I was very pleased by the process and the outcome. One of the results was the recommendation to introduce stringent controls and regulations. It was proposed, for example, that if you don't know for sure what kind of effect can be expected to be exerted from the genetically engineered organism, you should assume that it would be highly pathogenic and harmful. Once you have shown under appropriate laboratory conditions that it isn't harmful, then you can relax the precautions and go to less stringent conditions for further explorations, and eventually, for its lasting application.

Nature herself has been a genetic engineer.

All the time, yes. I have used this argument for several decades. I have indeed come to the conclusion that genetic engineering uses precisely the same strategies to alter genetic information as nature does to produce genetic variants for biological evolution. In the natural reality there are three different strategies to generate genetic variants. One strategy is changing single nucleotides or just a few; these are local changes in the DNA sequences. Another strategy is to rearrange some segments of DNA within the genome of the organism. The third strategy is to acquire a segment of DNA from another organism by horizontal gene transfer. Precisely these same three strategies serve in genetic engineering. The problem that I still face today is to convince my colleagues in biology, even in evolutionary biology, because they cannot understand these molecular processes, mainly since these are and must be quite rare events. Therefore, it will take quite some time until the scientific community of biologists will take it as a given and can go to politicians and to the general public to explain them how natural reality functions. That's the way scientific progress goes. It does not suffice that a scientist has his own conviction and interpretation of experimental data. It is only upon wide acceptance by the peers, the community of experts in the specific fields, that novel interpretations can become part of the scientific knowledge base.

Often, we agree about the data. However, since genetic variation is a very rare event, and since it is not and should not be reproducible from event to event, it's as a matter of fact only statistically reproducible; therefore, the interpretation of the natural strategies is not straightforward for people. If you read textbooks on genetics and on evolution, they give the impression that genetic changes are always due to accidents and errors in the genetic material. I feel that this is a wrong understanding of evolution. Evolution is rather something active, and it is relatively careful with the genetic information. All the species that live on our planet have a certain genetic stability. However, in the large populations, some individuals may suffer some kind of genetic change, each of a different type, at a different location in their genome. If the organism accepts such a change, the new genetic variant will be maintained and sometimes favored as compared with its parents, and it may eventually overgrow the parental population. That's the natural selection that's working there."[3]

Arber had quite a shocking message in conclusion.

"Some people are afraid of the development of science. They fear that applications of science will ultimately lead to the destruction of life on Earth. Maybe they are right to some degree. For me, biological evolution is an absolutely wonderful thing on this planet. There may be biological evolution elsewhere in the universe as well, but we can so far only evaluate what happens on our planet. I'm convinced that if living conditions would change so drastically that human life and the lives of some other highly developed organisms would no longer be feasible, life would still continue for a very long time because other organisms could exist under very different conditions. That makes me an optimist. I should add that I am not so much anthropocentric to say that for me only the human life form is what counts. For me, life as such is more important than the specific existence of human life. Therefore I have a big hope for the continuation of life on our planet as long as some living conditions continue to exist. We are just one life form among many and if some other organisms would survive, even if they would be very simple organisms, they could have a chance to develop into higher complexity again by the given means of biological evolution."[4]

---

[3] *Candid Science VI*, 154–57.
[4] Ibid., 163.

DAVID BALTIMORE

If you cannot have faith in your colleagues, then you can't do science, certainly not collaborative science.

Baltimore in 2004 at Caltech. (photo by I. Hargittai)

David Baltimore[1] (1938–) was born in New York City. He is a biomedical scientist. He received his bachelor of science degree from Swarthmore College in 1960 and his PhD degree from Rockefeller University in 1964. Following postdoctoral work, he was at the Massachusetts Institute of Technology from 1968 to 1990, at Rockefeller University from 1990 to 1994, and then at MIT again before moving to the California Institute of Technology. He had a series of prominent administrative positions in the Northeast. Then, between 1997 and 2006, he served as President of Caltech, where he now is President Emeritus. In 1975, he shared the Nobel Prize in Physiology or Medicine with Renato Dulbecco and Howard M. Temin "for their discoveries concerning the interaction between tumor viruses and the genetic material of the cell."[2] We recorded our conversation in 2004, in Baltimore's office at Caltech.

---

[1] I. Hargittai and M. Hargittai, *Candid Science VI*, 164–81.
[2] "The Nobel Prize in Physiology or Medicine," *Nobelprize.org*. Nobel Media AB 2013, http://www.nobelprize.org/nobel_prizes/medicine/laureates/1975/.

Browsing through Baltimore's statements about the role of government, we could detect a shift in his views.

"I am more conservative today. I no longer believe that government is the right solution for everything.

You used to believe that?

That's basically the socialist notion and taking it to the extreme; it's the communist notion. I give you an example, where I am absolutely convinced that it is not the right answer. That is when comparing the educational systems in the United States and Europe—I mean higher education. In Europe, all higher education is a state function. All research is a state function. The notion of a private university is very rare. It's coming, but it's still very rare. It's clear that the strength of the American higher educational system comes from the fact that we have a lot of private universities. Even the public universities are largely privately funded. There is no state university in the United States which gets more than 50 percent of its budget from the state. Most of them are in the range of 20 or 30 percent. There is a tremendous power and energy in private activities.

When recombinant DNA was first discovered, I said to a friend of mine that this was a clear case of when the government should take over the technology and to develop it in the interest of people. It should not be allowed to be a source of profits to individual companies. It was a great opportunity to do that because it was new, the government funded the research, and they could have easily stepped in and taken it over. My friends, who were in the real world—I was a professor at MIT at that time—they patted me on the head and told me that the only way to develop it was through private enterprise. I learned a huge lesson through that because it was correct; the government couldn't have done that effectively; and the market place drove it, and it drove us to enormous successes in the use of modern biology in the interest of making new pharmaceutical agents.

The lack here is anybody working in the public interest, and that's a huge lack. I would love to see a government agency and an international agency try to use modern biology in the interest particularly of the less-developed world. But I don't think that the leading edge of this business should be managed by government agencies. So, yes, my views about how society should be organized have changed, very much in parallel with the general changes in this country. Whereas in the days of Roosevelt, socialist notions had a lot of cachet, they have very little cachet today anywhere in the United States."[3]

At the conclusion of our conversation, Baltimore had a message.

"The message that I take from what I have seen of the world over the last 50 years is that the actual doings of the world are extremely messy, largely because they are so mired in politics. But if you take a very long baseline, and you look at the way the world is

---

[3] *Candid Science VI*, 171–72.

going, rational considerations, humane thinking, wins out. I watched the country I was born into, which was viciously segregated, which was totally opposed to many different groups, blacks, Jews, anything foreign to basic WASP society, evolve, over my lifetime, into a country that, if not totally accepting the differences, is much, much more accepting them. It is much, much more judging people as individuals than as members of groups, to the point when Massachusetts yesterday said that...separate facilities are intrinsically unequal. They have carried that as far as to say that a separate code of marriage for homosexuals is not equal; it does not provide equal protection under the law. That's an enormous distance for a country to go. With all the ups and downs of democracy, to see that we've actually run to a point like that is incredible. It's not over yet, and it may be a dominant topic for the upcoming elections, but it indicates which way the country is going."[4]

[4] Ibid., 180–81.

SEYMOUR BENZER

Max [Delbrück] was always a challenge.

Seymour Benzer in 2004 at Caltech. (photo by I. Hargittai)

Seymour Benzer[1] (1921–2007) was born in New York City. He was a physicist turned biologist. He received his bachelor of science degree from Brooklyn College in 1942 and his master of science and PhD degrees, in 1943 and 1947, respectively, from Purdue University, all in physics. Eventually, he changed from physics to biology, and he was already a biologist when he moved to the California Institute of Technology in Pasadena, where he stayed for the rest of his life. He was one of the pioneers of molecular biology. He is most famous for his discoveries of the relationship between genes and behavior and between genes and longevity. He was a much decorated scientist; one of his awards was the Crafoord Prize (Sweden, 1993). We recorded our conversation in 2004 in Benzer's office at Caltech.

---

[1] I. Hargittai and M. Hargittai, *Candid Science VI*, pp. 114–33.

Benzer had extensive postdoctoral experience working with world-renowned scientists, including Max Delbrück, who was a most influential scientist, albeit not for his discoveries.

"His influence went beyond that. He created a whole school of thought. In my case, Luria suggested that I ask for a copy of three lectures Delbrück had given at Vanderbilt, where he was then in the physics department. One lecture was precisely about the complementarity principle, and that was quite fascinating to me. The others were about the methods he had designed for quantitating work on bacteriophage. I had naively thought that a bacteriophage was the closest thing to a naked gene, and if you wanted to duplicate the gene, that was the best system to work with. What Delbrück taught us was how to do the experiments in a stepwise, quantitative way, which was very appealing to a physicist, in contrast to typical biology, which was descriptive. This was a system that you could analyze. It was also a mystery; you start with a bacteriophage particle that attacks the bacterium, and twenty minutes later a hundred descendents come out. What was happening in between? It was a tremendous challenge. That was what I tried to work on when I started in biology, using ultraviolet radiation to penetrate the cell, to analyze what was happening during this latent period.

Although Delbrück had tremendous stature, it was legendary how many times he could be wrong. But, as Jim Watson once told me, it was always for an interesting reason—in contrast to some other people. Max was always a challenge. Gunther Stent summed it up very well when he said that Delbrück was 'a conscience and a goad.' He set a standard for honesty, openness, and objectivity; but at the same time, he stimulated you by being skeptical. Even after you discovered something, he would say, 'No, I don't believe it.' One famous example was when my student Ron Konopka and I discovered mutations in flies that change their sense of time, the rhythm of their daily lives. I told Max [Delbrück] about it and he said, 'I don't believe it, that's impossible.' Even when I told him that we had already done it and we had found the gene, he only repeated, never mind, that's impossible. I must tell you that that was not an unusual event. Yet, somehow, he was a higher-level person, an inspiration to the people around him, and I was very fond of him and his spirited wife, Manny. Working in his lab at Caltech brought me in touch with wonderful colleagues like Renato Dulbecco, Jean Weigle, Gunther Stent, Elie Wollman and others."[2]

[2] Ibid., 120–21.

PAUL BERG

I don't know anybody who denies [a close relationship between genes and intelligence].

Paul Berg in 1999 at Stanford University. (photo by I. Hargittai)

Paul Berg[1] (1926–) was born in Brooklyn, New York. He received his PhD from Western Reserve University in 1952. He did postdoctoral work in Copenhagen and St. Louis, Missouri, and joined Washington University in 1955. In 1959, he moved to Stanford University and has been there ever since. He was awarded half of the 1980 Nobel Prize in Chemistry "for his fundamental studies of the biochemistry of nucleic acids, with particular regard to recombinant-DNA."[2] In addition to research, he has been involved in science policy and also in biotechnology companies. We recorded our conversation in 1999 in Berg's office at the Stanford University School of Medicine.

---

[1] Istvan Hargittai, (Ed. Magdolna Hargittai), *Candid Science II: Conversations with Famous Biomedical Scientists* (London: Imperial College Press, 2002), 154–81
[2] "The Nobel Prize in Chemistry 1980," *Nobelprize.org*. Nobel Media AB 2013, http://www.nobelprize.org/nobel_prizes/chemistry/laureates/1980/.

Decades before our conversation, when genetic-manipulation research was in its infancy, Berg had warned the scientific community that such experiments might represent potential biohazards for humans and the environment.

"When we began to do these experiments, people worried about, 'Would we be making bacteria more infectious?' In other words, if I take a gene from a cancer virus and I put it into a bacterium, and that bacterium can live in your intestine, does that create a big risk that you'll develop intestinal cancer? We didn't know the answer. If you take genes from organisms that are not sensitive to certain antibiotics and you put them into other bacteria that are sensitive to the antibiotics, do you change them to become resistant to the antibiotics? The answer to all those questions is yes. So at that point we said, 'Wait a minute, perhaps we should stop doing these kinds of experiments and determine if there is a real hazard or risk in doing these kinds of experiments.' We certainly did not want to send bacteria out into the environment that could infect people, cause cancer or whatever. At the time, however, we had no way to determine whether the risk was real or imaginary.

As soon as it became clear from these experiments that you could move genes to and from all kinds of organisms, people said, 'Are we going to create some monsters?' I was asked by the National Academy of Sciences to convene a group that would give advice to the Academy of what to do with this new science. This small group published a letter in the journals *Science* and *Nature*, which became known as the 'moratorium letter.' The 'moratorium letter' suggested that everybody around the world should stop doing these kinds of experiments until we could meet and determine whether they posed a risk or not. This letter had a profound effect all over the world, and every country issued it to all its scientists working in this field, 'Stop doing these experiments.' In that letter, we recommended that there be an international conference where the scientists working in this field would come together and try to arrive at a judgment about whether these experiments were risky.

As a result, a now very famous conference called the Asilomar Conference was held in 1975. Asilomar is a big conference center on the Monterey Peninsula near Carmel, about a hundred miles south from here. The conference brought together 150 people from all over the world—from Russia, China, Europe, and the United States—all people who wanted to use this technology of genetic manipulation. The purpose was to ask the question, is it safe? And if it's not safe, what can we do to reduce the risks so that we can do the research. At the conference, we decided that we didn't have enough information to determine whether it was really safe or not. What we could say was that there are some experiments for which there are almost zero risk; and here is a group of experiments for which there are possible risks, and another series of experiments for which we think there's serious risk. We recommended working out a set of instructions and guidelines which everybody would accept: These kinds of experiments could be done in any laboratory, these kinds of experiments would be done only in special laboratories, and these kinds of experiments would be done in only the most secure laboratories. The guidelines essentially regulated research for about twelve years. And because they were not set up by legislation, they were only recommendations; they could be changed.

As we learned more and more and found out that experiments we were worried about were perfectly safe, we could move them from the high-risk category to the low-risk category. Today, there's almost no restriction on the kinds of experiments and molecules that you can work on anywhere. The work was going slowly at the beginning because people had to build special laboratories for it; but as it became clear that it was safe, the work expanded and exploded. Today I am quite confident that there is no risk to any of the kinds of experiments that we have been doing over the last twenty years."[3]

On mapping the relationship between genes and behavior.

"Yes. It turns out that some behavioral traits stem from overproduction of a chemical. After all, look at what drug addiction does to your behavior. It's induced by the fact that a chemical comes in and acts on various cells in the brain to produce a complex readout, if you will. We don't know how many genes are involved in such a behavioral change. What we now know, because these experiments are going on, is that you can knock out one gene in a mouse and the mouse will fail to learn, can't learn, can't remember. In any kind of a training program, it will act as a stupid mouse. Another mutation can make the mouse a genius. It learns how to do the trick that you want it to learn in one try, whereas a normal mouse learns it in five to eight tries. All you've changed is one gene.

This points to a close relationship between genes and intelligence.

I don't know anybody who denies that. But it depends on how you define intelligence.

Isn't this almost a taboo question?

No, the experiments I mentioned are not concerned with humans. We're merely trying to understand what memory is. In what form is memory stored in our brains. Why do even very old people who don't remember what they did five minutes ago, remember the most minute details of various events from their youth? One way to learn about this is to ask if we can disrupt long-term memory. I will teach you how to do something, and then I'll challenge you a month later. If you remember—then what you had learned, that's long-term memory. But if you cannot learn, and I know what gene I had changed, I can say at least that that has something to do with the transformation from short-term to long-term memory. That's what people are trying to do. It turns out that the predominant paradigm in biology today in studying any complex phenomenon—it could be metabolism, it could be memory, it could be differentiation, it could be anything that you'd like to understand—is to make mutations that affect that process. Once you have collected all kinds of mutations that alter the process, then the challenge is to study what each mutation is doing and reconstruct all the steps that are responsible for this complex process. For example, certain mutant flies can't learn, or they change their behavior to light. Any kind of phenomenon that you want to study can be studied by making mutations.

Is there a one-to-one correspondence, or are there other changes too that people may neglect to follow? You make something live longer but become also stupid or genius.

That's a good question."[4]

---

[3] *Candid Science II*, 166–68.
[4] Ibid., 172–73.

BARUCH S. BLUMBERG

In my stage you don't change careers—you add them.

Baruch Blumberg in 2002 in Philadelphia. (photo by I. Hargittai)

Baruch S. Blumberg[1] (1925–2011) was born in New York City. He received his bachelor of science, doctor of medicine, and PhD degrees from, respectively, Union College in Schenectady, New York (1946); Columbia University (1951); and Oxford University, UK (1957). His most famous discovery was the hepatitis B virus and the creation of the vaccine against it. Blumberg shared the 1976 Nobel Prize in Physiology or Medicine with D. Carleton Gajdusek "for their discoveries concerning new mechanisms for the origin and dissemination of infectious diseases."[2] We recorded our conversation in 2002 at the Fox Chase Cancer Center in Philadelphia.

---

[1] Balazs Hargittai and Istvan Hargittai, *Candid Science V: Conversations with Famous Scientists* (London: Imperial College Press, 2005), 578–87.
[2] "The Nobel Prize in Physiology or Medicine 1976," *Nobelprize.org*. Nobel Media AB 2013, http://www.nobelprize.org/nobel_prizes/medicine/laureates/1976/.

Blumberg's office at the Fox Chase Cancer Center was in a building dedicated to the prevention of cancer. In view of this and his seminal work in preventing liver cancer, the question about cancer prevention was quite natural.

"The major program for the prevention of cancer has been the program for the cessation of smoking. That has had a profound effect on the incidence of prevalence of cancer, primarily in the United States. My work is connected with the prevention of cancer through the hepatitis B vaccination program. Hepatitis B accounts for about 85 percent of the primary cancer of the liver in the world. Primary cancer of the liver is one of the most common cancers; it's the third most common cause of death from cancer in males and the seventh most common cause of death from cancer in females. It's very difficult to treat. The life expectancy after clinical diagnosis is much less than a year, and the survival rate for five years—a frequent method of measuring the severity of a cancer—is 8 to 10 percent, which is extremely low. Another major cause of primary cancer of the liver is hepatitis C, either by itself or in combination with hepatitis B.

We invented a vaccine in 1969 to prevent infection with hepatitis B. That vaccine became available for general distribution in the 1980s. It is now one of the most commonly used vaccines in the world. More than a billion doses have been used; hundreds of millions of people have been vaccinated. This led to a striking decrease in the prevalence of hepatitis B infections in the world. Prevention of hepatitis B is the second most common intervention program for prevention of cancer. Cessation of cigarette smoking is the first."[3]

In the late 1990s, Blumberg embarked on a new career when he became the founding director of the Astrobiological Institute of the National Aeronautics and Space Administration (NASA). It was interesting to learn what, if anything, besides having earned a Nobel Prize (in a different field) might have justified his appointment.

"Strangely—what I had not realized—it was my skills at management, in managing a research organization. The Astrobiological Institute is a basic science institution. In this, it is different from other parts of NASA, which is a mission-driven organization. NASA is primarily an engineering and technology operation. In order to do any science, in order to get into space, you have to build these incredible spaceships and stations. The International Space Station is probably the most complicated thing that has ever been built; it is more complicated than the pyramids, the highway system, or the railroad network. It includes all kinds of pioneering engineering skills. It's a remarkable piece of work—amazing that it's there. However, the purpose of all this is to do science. Space is a great mystery. Every time we look at something, observe something, it is not only that nobody has ever been there before—nobody *could* have ever been there before. Everything is new. It's like when humans used the telescope for the first time. Every time they looked through the tube they discovered something new.

---

[3] *Candid Science V*, 580.

Do you expect to learn something about the evolution of life on Earth from your studies?

A lot of our work is focused on that. We're particularly interested in early evolution, the very start of life, and prebiotic chemistry. There is a tremendous amount of organic molecules of several hundred compounds that falls on Earth every year in form of meteorites and other space dust.

Might that be the origin of life on Earth?

That is a possibility."[4]

---

[4] Ibid., 585–86.

SYDNEY BRENNER

What science depends on is taking not the "morest" data but the "leastest" best and predicting the remainder from some other information.

Sydney Brenner in 2003 in Cambridge, England. (photo by I. Hargittai)

Sydney Brenner[1] (1927–) was born in Germiston, South Africa. He received his first degree from the University of Witwatersrand in South Africa; and in 1954, his PhD degree from Oxford University, UK. In 2002, Brenner shared the Nobel Prize in Physiology or Medicine with H. Robert Horvitz and John E. Sulston "for their discoveries concerning genetic regulation of organ development and programmed cell death."[2] By then, Brenner had received most other possible awards for his scientific achievements. One of his current affiliations is the Salk Institute for Biological Studies in La Jolla, California. We recorded our conversation in 2003, at King's College in Cambridge, England.

---

[1] I. Hargittai and M. Hargittai, *Candid Science VI*, 20–39.
[2] "The Nobel Prize in Physiology or Medicine 2002," *Nobelprize.org*. Nobel Media AB 2013, http://www.nobelprize.org/nobel_prizes/medicine/laureates/2002/.

On whether it is possible to compute organisms from their DNA.

"All we know is this linear script—and we have known it for some time—and we know that regions of it are translated into an amino acid sequence. We also know that some other regions carry information for products which are themselves nucleic acids. We know about transfer RNA, and we know about lots of other small RNAs, which have been discovered, but which we don't know so much about. Then, of course, we know about the ribosome and several other entities in the cell. We also know that in some way the regulation is written there. But we don't have a lexicon and we can't interpret it. So if you were to ask, can we compute organisms from their DNA, the answer is no. Not now."[3]

On explaining complex things.

"John von Neumann said that there were two ways of explaining complex things. One was to explain them in terms of essentially the level above, that is, in a matter language, in other words. Then he said, certain things are so complex that effectively we cannot...explain them. And we had to define a prescription for constructing objects that behave the same way; in other words, to give an algorithmic explanation of them. He accepted that as a scientific explanation. As an example he quoted pattern recognition. Now, how to explain pattern recognition itself; what you can do is describe the essential features of pattern recognition, to describe an object with this internal structure. Usually now it is a computer program rather than just the solution. So the answer is, I believe, the following: we can describe everything in the universe today; we have the power to give, atom-by-atom, a description of everything. But that's just data; that's just description. What science depends on is taking, not the 'morest' data but the 'leastest' best, and predicting the remainder from some other information. In other words, it is the classic technique of science to effectively form a theory from the facts as ascertained, and then you can predict it. For explaining such things as the behavior of organisms, we could essentially make a description of how an organism behaves under all circumstances, but that is description. The best thing, I believe, is to know what generates the behavior, the machine, the structure, and then we can predict the behavior. Once you have that you have the explanation of it."[4]

On Erwin Schrödinger's book What Is Life?

"There is a terrible error in Schrödinger's book, which is a fundamental mistake. There is a section in which he states that the chromosomes contain the plan or the program of the organism and the means to execute it. They do not contain the means to execute it. They contain a description of the means to execute it. That's a fundamental error, which I saw only later when I read von Neumann's theory on self-reproducing machines. The code in his model does not contain the means to execute the program, it only contains a description and you have to use the old machine to make the new machine.

[3] *Candid Science VI*, 24–25.
[4] Ibid., 25.

Schrödinger went wrong there. So I read it, and in hindsight, in going through it again, that is a fundamental error. My copy of *What Is Life* has a quotation in it from Michael Faraday, which I penciled in when I read the book in 1946 and which I don't quite remember now, but its essence is that you go out and do experiments. This is important and especially in biology where a theory is a theory, so what?"[5]

---

[5] Ibid., 27–28.

ARVID CARLSSON

If it weren't for him[Bernard Brodie], these discoveries wouldn't have happened. He was a great pioneer of modern pharmacology.

Arvid Carlsson in 2003 in Gothenburg, Sweden. (photo by I. Hargittai)

Arvid Carlsson[1] (1923–) was born in Uppsala, Sweden, and began his studies in 1941 at the University of Lund, He received his Swedish M.L. degree (corresponding to the American MD) and his Swedish M.D. degree (corresponding to the American PhD), both in 1951. He has been at the University of Gothenburg since 1959. Early in his career, in 1955–56, he spent half a year working with Bernard B. Brodie at the National Heart Institute in Bethesda, Maryland.

Carlsson shared the 2000 Nobel Prize in Physiology or Medicine with Paul Greengard and Eric Kandel "for their discoveries concerning signal transduction in the nervous system."[2] In 1958, Carlsson and his colleagues identified dopamine in the brain and proposed its agonist function in the control

---

[1] B. Hargittai and I. Hargittai, *Candid Science V*, 588–617.
[2] "The Nobel Prize in Physiology or Medicine 2000," *Nobelprize.org*. Nobel Media AB 2013, http://www.nobelprize.org/nobel_prizes/medicine/laureates/2000/.

Bernard B. Brodie. (courtesy of Arvid Carlsson)

of psychomotor activity. He and his students discovered the distribution of dopamine in the brain and proposed a role for it in Parkinson's disease. He has made numerous other discoveries in pharmacology, as well. We recorded our conversation in 2003 at the University of Gothenburg.

Carlsson had written somewhere that he and Brodie had the advantage of being outsiders. We asked him what he had meant by that.

"We were outsiders and that was very important. This fits well with [Thomas] Kuhn's teachings. He describes the solid establishment, the dogma. Then somebody comes in from the outside and says, it can't be true. One of the things that impressed me about Brodie was that he could listen to a seminar outside of his field, and at the end, he would come up with some penetrating questions.

After about sixty years in science, you keep returning to those five months that you spent with Brodie.

He was fascinating in many different ways. I still think about him, about what really made him such a character. Let me give you a couple of examples that make it even more mysterious. The discovery of serotonin depletion by reserpine was truly groundbreaking. But that was not his discovery; he just picked it up. The discovery itself was made by a young man in his lab, Parkhurst Shore, who, by the way, was my mentor, under Brodie's supervision, during my stay in Brodie's lab. I had a phone conversation with Shore just a couple of weeks ago because I wanted to find out a little bit more about the thing. It's mentioned in the book *Apprentice to Genius* that it was Shore. Shore was put on a problem that had to do with a Brodie kind of concept, that is, how do two drugs interact with each other? It was a basic problem. In this context, they were investigating also drugs that acted on the central nervous system (CNS). Shore started to read about this and he was fascinated by the finding that there are a number of CNS agents that were indoles, serotonin, and reserpine, and LSD. Then, it so happened that there was a younger collaborator of Udenfriend's, Herb Weissbach, who had developed a method to measure the metabolite of serotonin in the urine. Shore went to Weissbach and asked him if he would like to analyze urine from animals that had been given reserpine. So they did that, and they were astounded by the results. They had never seen such an elevation in the metabolite level. At the time Brodie was on vacation in Florida, but as soon as he heard about this, he rushed back and immediately started talking about 'our project.' He took over. Shore is such a modest man. He says, 'That's just how the system works.'

Shore had another discovery. If you give certain drugs that are N-containing bases and then analyze the content of the stomach, you see an enrichment of the bases there. From this, they could derive the idea that it's a matter of distribution between two phases in the stomach, and if it is a base, it will be enriched in the acid phase. That was also Shore's finding, but Brodie gave it over to another guy who was more of a specialist of the stomach in Brodie's group. There was a publication from it with Shore's name as first author. It was the structure of Brodie's lab.

Then it was Julie [Julius] Axelrod who went to a neighboring lab and learned to prepare and fractionate liver. He discovered the role of the microsomes in drug metabolism. It also became Brodie's discovery, not only Axelrod's. There were even jokes about it, about Brodie discovering the effect of reserpine on serotonin on the beach in Miami. These are negative aspects of Brodie. But he was the one who created this atmosphere of creativity; he did not develop the spectrophotofluorometer, but it was created in the atmosphere of his lab, and he may well have had the original idea. The spectrophotofluorometer was a marvelous instrument in those days which opened up new possibilities; for the first time, it became possible to measure small amounts of many important substances with chemical methods.

All this happened in Brodie's lab. He attracted these people who did all this. So, what can we say? He was a pioneer of a field. Even though he was a greedy kind of pioneer, he was a pioneer. If it weren't for him, these discoveries wouldn't have happened. He was a great pioneer of modern pharmacology. But taken everything together, there could also be found arguments against a Nobel Prize for him.[3]

Your discovery provided a cure or alleviation for Parkinson's syndrome.

Yes. I formulated the whole concept of dopamine's role in Parkinson's disease. That was in October 1958 in Bethesda, Maryland.

Why do we still see so many people suffering from Parkinson's disease?

This is not a matter of prevention; it is a matter of treatment, a symptomatic treatment. You replace what has been lost through cell death by dopamine, by giving its precursor DOPA. In the majority of patients, there is a dramatic improvement, which can last for even more than a decade. In some cases, and sooner or later in most cases, the therapeutic action of L-DOPA is weakened. Side effects show up, such as the so-called on-off-phenomena, so that the patient is Parkinsonian stiff, and after various intervals, all of a sudden switching, and then is moving freely. Then, all of a sudden, the patient

---

[3] *Candid Science V*, 613–15.

switches off again and is back to this stiff, motionless state. There is another side effect, the so-called dyskinesia, involuntary movements. It can be very disturbing.

When you say that it is treatment, and not prevention, this resembles the insulin story, with a twist. Following the Nobel Prize for the discovery of insulin, there was criticism that the Nobel award diverted attention from diabetes research to find the cause and a cure. Might the recognition of dopamine dampen further research on Parkinson's disease?

My answer would be that it's exactly the opposite. Parkinson's disease didn't attract any interest by science at all. It was a chronic disease; it was due to some cell degeneration, but the cells that degenerated were not even identified. Obviously, there was some neurodegeneration. What happened with these poor patients was that the diagnosis was established, and the doctor said, 'This is a chronic disorder, I'm sorry, we can't do anything about it.' There was very little scientific interest in this. After my discovery, Parkinson's disease became a subject of guided brain research, both basic and clinical research. It opened up a new concept, that even if you have a degenerative disorder, it's possible to treat it. That was something new. But there I also met some resistance. This time it was from the neurologists. They thought that this was a ridiculous idea—that is, if you have a neurodegeneration, to introduce a chemical. They thought it was nonsense. Even here [Gothenburg]. When I came here in 1959, I immediately contacted the professor of neurology and suggested a collaboration. He was not at all interested. I did the same thing with a famous neurosurgeon in Lund, where I was before I moved here, and he was not interested either. To them this idea was impossible because in their minds, they had a model for the brain where one cell communicated with another by an electric spark, so how could a chemical do anything of interest in this context. Therefore, this discovery was not only a paradigm shift from the point of view of basic research, it was a paradigm shift in clinical neurology, and even psychiatry, for that matter. It had a tremendous impact, exactly the opposite of what you proposed.

I didn't really propose it.

I know, of course [laughing heartily]. You just wanted to tickle me a little bit."[4]

---

[4] Ibid., 596--97.

AARON CIECHANOVER

Israel is a miracle in my eyes.

Aaron Ciechanover in 2003 on Sandhamn, an island in the Stockholm Archipelago.
(photo by I. Hargittai)

Aaron Ciechanover[1] (1947–) was born in Haifa. He is an Israeli biomedical scientist who received both his master of science and doctor of medicine degrees from the Hebrew University School of Medicine in Jerusalem. He has been with the Technion–Israel Institute of Technology since 1977. He has had numerous visiting appointments, including one from 1978 to 1981 at the Fox Chase Cancer Center in Philadelphia. Ciechanover was co-recipient of the 2004 Nobel Prize in Chemistry together with his former mentor Avram Hershko and Irwin A. Rose "for the discovery of ubiquitin-mediated protein degradation."[2] We recorded two long conversations. The first took place in 2003 in Sweden on the island of Sandhamn in the Stockholm Archipelago, during the second annual retreat for cancer researchers of the Karolinska Institute. The second conversation was recorded in 2005 in Budapest.

---

[1] I. Hargittai and M. Hargittai, *Candid Science VI*, 258–303.
[2] "The Nobel Prize in Chemistry 2004," *Nobelprize.org*. Nobel Media AB 2013, http://www.nobelprize.org/nobel_prizes/chemistry/laureates/2004/.

The following is part of Ciechanover's response to a question from a student at the Lauder Jewish School in Budapest about how he reconciled religion and science.

"I am not facing such a problem, because I don't believe that I am a religious person. I am very Jewish, but I am not a religious Jew. There must be a distinction.... For me, Judaism is something completely else. I am very keen on Judaism.... For me, Judaism is tradition, culture, history, remembrance. It's parents and home, scholarship. It's many things. I don't feel that I have to reconcile science and worshipping something that I do not understand and that is above me.

Judaism for me is a real, deep value that appears in a great variety of appearance. Let's take my hobby.... There is a huge difference in the sense that the Jewish cantorial music carries the prayers and the suffering of Jews along centuries, the conversations between them and their God. Some of them are written in an extremely strong and poetic way.

Again, it's not the God in the sense of a Jewish God. It's the God in the sense that places us, human beings, in the right place, where we belong, in terms of humbleness in appreciating creation, appreciating nature, the complexity of biological sciences. We need to be very humble because the world is so complex that we don't understand it. You can get a Nobel Prize [for discoveries] on the brain, but we don't understand the brain. We can get the Nobel Prize in advancing some of our knowledge of cancer, but we don't have a clue what cancer is, or how to cure it.

I give just one example. It takes me back to my father. There is a very famous prayer before the Musaf prayer. The Musaf prayer is a very important prayer, and it takes a real turn of importance during High Holidays, during the Rosh Hashannah (first day of the year) and Yom Kippur (the Day of Atonement). In this prayer, the Shaliah Tsibur, the Chazan (the cantor, the representative of the community) sings the prayer for the community. He stands before his God, before the Ark, and says, in my translation, 'I am nothing. I am a piece of dust and I am worth nothing. I am just a simple human being. But I have a responsibility. I was sent by this community, thousands of people, that are gathering in this synagogue to pray for you, that you will forgive them. I am just their messenger. So, please open the gate of heaven for me, because it's not for me, it's for them and for everybody.'

I shiver when I read this, and it doesn't matter which God, it can be any God.... How did the Jews live? They lived a religious life. They wrote their codex of laws, the Talmud and the Mishnah. What is the Talmud with the Mishnah?—it is a civilian law codex. What happens if I rob you, what happens if I rape your wife, what happens if I violate this and that? They assembled it into a way telling everybody how a community should live. This has become a religious script, but it is basically a civilian script. It is one of the oldest codes of laws in the world. So, obviously, I appreciate it, it is Jewish, but it has nothing to do with God. It is how human beings should live together in a community. Obviously, there are some laws of sacrifice in Jerusalem. You go to the temple on the High Holiday, and you sacrifice to your God and bring part of your harvest, [your] agricultural products, to God, but you give it also to the poor. On every field you harvest, you always keep the corner of the field not harvested so that the poor people can come and take [from] it, because they don't have it. There are many social rules built into it. This is all in the Talmud and the Mishnah, which are religious scripts, but these are the ways we should live.

Then. it obviously has to do with my parents. My parents grew up in Poland, and then they made it to Israel before the Holocaust and they grew up orthodox. They left me the tradition. This is the tradition part, and now we are coming to the State of Israel. The State of Israel was recognized by the United Nations on November 29, 1947, and was established as Ben Gurion's independent state on May 15, 1948, as the direct result of the Holocaust. It wasn't established in the previous century, nor was it established fifty years after the Holocaust. It was established right after the Holocaust. And what is the Holocaust? It is the destruction of Jews *because* they are Jews. They were cruelly murdered because of it; persecuted all over Europe; and murdered by the Germans. So, how can I *not* relate to it? I *must* relate to it, I am part of it. I am living in a country that is the direct result of the Holocaust, a religious persecution. It was racial in the eyes of the Germans, but in my eyes, after many years of anti-Semitism and persecution, [the Holocaust] was just the end of it, the culmination.

Then Jews established their own state and established their own army, with two purposes. One. that they would be able to protect themselves, doing something that nobody can do for them. And, to bring Jews from all over the world and tell them: if you are a Jew, you are entitled to be here and this is your shelter; you are not asked any questions. This is the only state in the world that gives automatic citizenship to Jews just because they are Jews. This is a shelter from a physical threat, not just from a religious threat; things are linked together. It has nothing to do with whether I get up in the morning and pray according to one book or another. It is a chain of history, the Talmud, the Mishnah, the persecution, the culture, and my parents, my state. It altogether establishes a very unique—in my eyes important for my very existence—structure that I live in. I don't tell you, there is no problem of reconciliation between science and religion, but I never faced it. I never made an excuse for it either, and I never felt that I needed to make an excuse for it in order to satisfy people who ask me 'How can you be a scientist?' There are scientists who wear a kipa and are really religious in the full sense of the word. They may find it difficult, and I am sure that they have thought of an explanation."[3]

---

[3] *Candid Science VI*, 293–95.

FRANCIS H. C. CRICK

I would stress the right of a person who is incurably ill to terminate his own life.

Francis H. C. Crick in 2004 in La Jolla. (photo by I. Hargittai)

Francis H. C. Crick[1] (1916–2004) was born in Northampton, England, and died in La Jolla, California. Crick attended Northampton Grammar School, then the Mill Hill School in North London. He studied physics at University College London and received his bachelor of science degree in 1937. World War II interrupted his doctoral studies while he did war-related research for the British Admiralty. In 1947, he started doing research for the Medical Research Council (MRC), and from 1949, he worked at the Cavendish Laboratory of Cambridge University.

His informal cooperation with James Watson started in 1951, and it led to the discovery of the double helix in 1953. Crick, along with Watson and Maurice Wilkins, were jointly awarded the 1962 Nobel Prize in Physiology or Medicine.[2] Crick subsequently continued working in molecular biology, being

---

[1] I. Hargittai and M. Hargittai, *Candid Science VI*, 2–19
[2] "The Nobel Prize in Physiology or Medicine 1962," *Nobelprize.org*, Nobel Media AB 2013. http://www.nobelprize.org/nobel_prizes/medicine/laureates/1962

Odile and Francis Crick with Magdolna Hargittai in 2004 in La Jolla, California.
(photo by I. Hargittai)

primarily involved in the understanding of protein synthesis and the genetic code. He left Great Britain and joined the Salk Institute in La Jolla in 1976, and changed his research fields to understanding the brain and the nature of consciousness.

We visited Crick and his wife at their home in La Jolla in February 2004, just a few months before he died. We spent a long morning and had a nice lunch with the Cricks, and our conversation covered a wealth of topics. Both Francis and Odile were in great spirits. We did not record our conversation, but immediately following the visit, we (IH and MH) each narrated everything into a dictaphone. The segment that follows is from our correspondence with Crick.

On October 24, 1968, Crick gave the prestigious UCL Rickman Godlee Lecture. He gave it an intriguing title, 'The Social Impact of Biology.' He talked about some highly charged issues in that lecture, and we asked him about them, particularly whether his opinion during the decades since the lecture might have changed. In the lecture, he had raised the possibility that a baby should only be declared alive two days after birth. Another suggestion was to stop spending resources on medical care for people above eighty or eight-five years of age. At the time of our exchange, Crick was well over eighty.

"I would indeed modify my suggestions today. In the old days, doctors quickly let a very deformed or handicapped baby die, rather than make exceptional efforts, as they often do now, to keep the baby alive. I now realize that it would be impossible, at least in this country, to count life as starting after the first two days of a baby's life because so many religious people believe life effectively starts much earlier, even at conception. In other words, one has to consider not just the feelings of the baby (who hardly has any) but also the feelings of the parents, and of other members of society however silly one may think them to be. But I do believe that doctors should not make exceptional efforts to keep a very handicapped baby alive.

As to the age limit, people now live longer than they did in the sixties, so I think such an age might be a little higher, but I doubt if a rigid rule would be acceptable. Again, I think very expensive treatments, or ones that have only a limited availability, should be allocated in some sensible way. I've heard that the State of Oregon is trying out such a scheme.

If I were to give such a lecture again (which is unlikely) I would instead stress the right of a person who is incurably ill to terminate his own life. I believe this is being tried out in Holland."[3]

To the question about whether there were any scientists that could be considered directly his pupils, Crick had this to say.

"I don't think there is anyone whom I could call my pupil. I only supervised a graduate student for a year, but after that year someone else took him over. I think I deliberately avoided such tasks.

On the other hand, I have had several close collaborators. The major ones have been Jim Watson, Sydney Brenner, and (more recently) Christof Koch. Others I have had more than transient collaborations with are Aaron Klug, Beatrice Magdoff, Leslie Orgel, and Graeme Mitchison. In all these collaborations, we have published papers together. These collaborators (except possibly for Magdoff and Mitchison) have each had many pupils of their own.

I think I work best, not entirely by myself, but with one other close collaborator. Sydney Brenner and I shared an office for twenty years. At the moment my close collaborator is Christof Koch, a neuroscientist at Caltech.

Of course, I have interacted for most of my scientific life with a very large number of scientists and over the years have given lectures in many different places. Some people have told me that they were strongly influenced by a lecture of mine they heard. I think I must have been rather a good lecturer, because at meetings no one liked to have to lecture after me!"[4]

---

[3] Ibid., 11
[4] Ibid., 12.

D. CARLETON GAJDUSEK

I find fun in science only when I am thought to be a charlatan.

D. Carleton Gajdusek in 1999 in Budapest. (photo by I. Hargittai)

D. Carleton Gajdusek[1] (1923–2008) was born in Yonkers, New York. He studied at the University of Rochester from 1940 to 1943. He earned his doctor of medicine degree from Harvard University in 1949. By then, he was also completing a postdoctoral stint at the California Institute of Technology with Linus Pauling, John Kirkwood, and Max Delbrück. In 1951, he was drafted to Walter Reed Army Medical Service Graduate School. He worked from 1957 at the National Institutes of Heath (NIH).

Gajdusek studied always-fatal subacute diseases of the nervous system, including the disease called kuru, found in people of stone-age culture in New Guinea, who at the time Gajdusek arrived there, still practiced cannibalism. He showed that kuru was transmissible and caused by a new type of infectious agent that spread when members of the tribe ate their dead relatives. He further showed that this infectious agent was closely related to the one that caused

---

[1] *Candid Science II*, pp. 442–465.

scrapie in sheep; that the rare presenile dementia Creutzfeldt-Jakob disease is caused by the same atypical, unconventional "virus"; and that bovine spongiform encephalopathy (mad cow disease) is caused by one of the same group of agents. Gajdusek and Baruch S. Blumberg shared the 1976 Nobel Prize in Physiology or Medicine "for their discoveries concerning new mechanisms for the origin and dissemination of infectious diseases."[2]

Gajdusek adopted, brought to the United States, and educated over sixty children from the stone-age cultures of New Guinea and Micronesia. Later, he lived in Europe after having gone through a painful period in the United States, which included arrest and a prison term following charges of child molestation and financial irregularities. A tremendous show of loyalty by most of his adopted children and many others, including famous colleagues, eased this hardship. He continued writing his famous journals, which meticulously recorded his life and copies of which are stored in archives and with friends in several corners of the world.

We recorded two long conversations with Gajdusek in Budapest in 1999.

He talked about what he considered to be his most important contributions.

"If I just list the fields in which I feel I made my major contributions, it is not for those for which I won the Nobel Prize. I think the most important field to which I've made my contribution is in the variation in cognitive function of the brain—variability in brain programming for a given task in modern society and in stone-age peoples. That's the study, which encompasses even my crystallography and molecular genetics, slow virus diseases and Alzheimer's and Creutzfeldt-Jakob disease laboratories. The basic name of my section at NIH for forty-seven years was Study of Child Growth, Development and Behavior and Disease Patterns in Primitive Cultures.

The next most important thing I did was that I investigated the fact that the Claude Bernard experimental method is the worst of all approaches to science when you're dealing with nonrecurring events. It is unreasonable to test any theory on a nonrecurring event. No matter what your results, somebody can say you cheated or it's a lie. So why do so? I published in *Nature*, back around 1959 or 1960, a thesis on the study of nonrecurring events. I felt embarrassed because the people that used that paper later were NASA astronomers. You don't wait hundreds of years for an unusual comet to come back or a mysterious cosmological event to reappear. Everything happening when I walk into a stone-age community that is thousands of historical years behind our civilization technologically is changed by my very approach when they see me. Everything that occurs is never to occur again. I was thinking about this in our work using sound and cinema recording and found the need for random scanning at human scale instead of directed hypothesis testing. Observations directed by hypothesis are

---

[2] "The Nobel Prize in Physiology or Medicine 1976," *Nobelprize.org*. Nobel Media AB 2013, http://www.nobelprize.org/nobel_prizes/medicine/laureates/1976/.

the least valuable observations. Just observe, scan, and gather data, and try not to observe from only one theoretical point of view.

I think those are perhaps the most important contributions I made in science, all of them back fifty years ago.

Back in the forties I worked with Michael Heidelberger at Columbia University College of Physicians and Surgeons on the autoimmune reactions. Then I left the field and I worked on tissue cultures; I first used tissue cultures for a whole bunch of viruses, including herpes simplex and Venezuelan equine encephalitis viruses. Then when I went with Burnet to the Walter and Eliza Hall Institute in Melbourne, Australia, I started to work on influenza virus genetics. I had wanted to work on infectious hepatitis virus, but Burnet didn't want me to work on it because it's a dangerous virus. Later, he let me work with it, and I found some peculiar autoimmune reactions in hepatitis patients. In developing the autoimmune complement fixation [AICF] test, I started the study of immunology at the Walter and Eliza Hall Institute; previously it was all virology. I published in *Nature* a set of papers on autoimmune antibodies to various tissue components in hepatitis infections and yellow fever, and in multiple myeloma patients. It was using that data that Burnet, on his return, changed the Institute's focus to immunology, but he credited me for that. I would never have stated explicitly that each patient with multiple myeloma had a different clone of antibody-producing cells. That was the indication of the clonal hypothesis of antibody production. Burnet was smarter than I in using my data.

I had worked earlier on arboviruses in South America, and I found a new subgroup of arboviruses that were killing people. Later, I had for many years the unique hantavirus laboratory in the US in which we isolated worldwide most of the first dozen hantavirus strains and characterized them serologically. Ho Wang Lee, the discoverer of hantaviruses, sent me his student Pyong Wuo Lee and others from his hantavirus laboratory in Korea. He and Pyong Wuo Lee had first isolated the virus causing Korean hemorrhagic fever. It was with the Koreans in my laboratory that we plotted out the world hantavirus distributions from China to Scandinavia and in the US, the Balkans, Australia, Oceania, and South America."[3]

On what kind of science he respected the most.

"I don't respect that part of modern science which has been made into a soccer game. I have too large an ego to play...games. If what you're doing will be done by somebody else in the next ten years if you don't do it, stop doing it. I don't call that creative science. I do a lot of that too, it's the bread and butter of PhD candidates and young postdocs, who *must* publish. We just have to do it, but I don't respect it. Riddle solving is not the intelligence I most admire. I don't think much of the intelligences of the grand masters of chess. It's competitive masturbation of intellect. I gave up chess; I don't like the whole idea.

---

[3] Ibid., 454–456.

I find fun in science only when I am thought to be a charlatan. You don't know what you are doing, you don't know what questions to ask. You mull it over and you have foolish ideas for ten or twenty years; you talk to your colleagues, and they don't get it and get bored. That is the creative process of science. When you know what questions to ask and how to approach them and can finally get your colleagues excited, and they run home to write a grant proposal, you know you've done your job, and you move elsewhere.

The biggest damage to diabetics research was Banting and Best's discovery of insulin. It caused fifty years in which little work was done on the cause or prevention of diabetes, only studies on physiology of insulin, production of different insulin pharmaceuticals, and desensitizing people who are sensitive to insulin. It has nothing to do with ever preventing or curing diabetes. The same with multiple sclerosis. Today we know no more about the cause than we did in the early twentieth century. The same is for schizophrenia. I am waiting for the eighteen-year-old to come into my office, saying, 'I'm going to give my life to find the cause of schizophrenia.'"[4]

---

[4] Ibid., 458.

WALTER GILBERT

The "clone" is like a delayed twin, but it has different mitochondria, a different womb, a different upbringing, and a different age.

Walter Gilbert and Frederick Sanger in 2001 in Stockholm. (photo by I. Hargittai)

Walter Gilbert[1] (1932–) is an American theoretical physicist turned molecular biologist. He was born in Boston, Massachusetts, and received his first degrees in physics at Harvard University. For his PhD in mathematics (theoretical physics) in 1954, he went to Cambridge University, UK. He joined Harvard University in 1957, first in theoretical physics, later working in molecular biology. He was also active in biotechnology companies. Gilbert shared half of the 1980 Nobel Prize in Chemistry with Frederick Sanger "for their contributions concerning the determination of base sequences in nucleic acids."[2] (Paul

---

[1] *Candid Science II*, pp. 98–113.
[2] "The Nobel Prize in Chemistry 1980," *Nobelprize.org*. Nobel Media AB 2013, http://www.nobelprize.org/nobel_prizes/chemistry/laureates/1980/.

Berg received the other half of the prize.) We recorded our conversation in 1998 during the meeting "Frontiers in Biomedical Research" in Indian Wells, California.

Gilbert had a most varied career, moving from theoretical physics to molecular biology and biotech companies, picking up a chemistry Nobel Prize on the way.

"My career may be a message about the usefulness of being educated broadly. The problem is knowing enough odd little things about the world. DNA sequencing depended on knowing, at a certain moment, a certain strange fact about the sugars in DNA. How did I happen to know that? It was probably some experiment I had been involved in four or five years before. It would be very hard to follow the complete connection through. Edison used to say that being an inventor involves knowing all kinds of apparently irrelevant connections about the world.

The other message is that it's easier to change fields than one thinks. What handicaps people is that they're trained to do certain things, and they think they can only do what they're trained to do. This is basically untrue. The liberal education that the British used to use had the great advantage that by being trained for nothing, one felt free to do anything. If you read the classics, they obviously don't train you to do anything particular, but it is a training for the world in a fundamental way. One of the effects of the genome project on biology is that all the ways in which we used to do biology, the cloning and sequencing of genes, are going to disappear. Such things will no longer be research projects; they will just be something to look up. When I first did experiments with messenger RNA, we counted the radioactivity with a hand counter. Then the techniques got all automated, but now people are no longer using radioactivity at all.

There is constant change in the sciences, and if you don't realize this, you remain frozen behind. When I changed from a mathematical kind of physics to experimental biology, I realized that in my work for my physics PhD, I learned an essential thing that was transferable. That particular ability was how to decide myself whether something was right or wrong. That was the crucial element, and that ability was transferable to another field. I had to learn the particular knowledge of my new field, but one can learn the particular knowledge of any field in a short time."[3]

---

[3] Ibid., 112–113.

AVRAM HERSHKO

I like to work by myself.

Avram Hershko in 2004 in Woods Hole, Massachusetts. (photo by I. Hargittai)

Avram Hershko[1] (1937–) was born in Karcag, Hungary. He was six years old when he, his brother, and their mother were incarcerated in a ghetto and then put on a freight train destined for Auschwitz. By a lucky accident, the train was diverted to Strasshof, Austria. After spending months in a labor camp in inhuman conditions, following liberation, they returned to Karcag where the father joined them, returning from a slave labor camp. Soon, the Hershko family moved to Budapest, and in 1950 they moved to Israel. Hershko received his doctor of medicine degree in 1965 and his PhD in 1969, both from the Hebrew University–Hadassah Medical School in Jerusalem. From 1969 to 1971, he spent a postdoctoral stint at the University of California Medical School in San Francisco. He has been at the Technion–Israel Institute of Technology since 1972. He used to spend his summers conducting research at the Fox Chase

[1] *Candid Science VI*, pp. 238–257.

From right to left, Avram Hershko, Magdolna Hargittai, Judy Hershko, and Istvan Hargittai in 2005 in Budapest. (photographer unknown)

Cancer Center in Philadelphia. More recently, he spends his summers in Woods Hole, Massachusetts.

Hershko shared the 2004 Nobel Prize in Chemistry with his former student Aaron Ciechanover and with his former host at the Fox Chase Cancer Center Irwin Rose "for the discovery of ubiquitin-mediated protein degradation."[2] We recorded our conversation in August 2004, just two months before the Nobel Prize announcement, in Woods Hole.

One of us (IH) had a childhood experience similar to that of Hershko—the two might have even met in Strasshof, so it was natural to ask him about how he survived the Holocaust.

"First we were in a ghetto in Karcag, and I have some faint recollections of the gendarmes coming and ordering us out of our home. In the ghetto, my mother tried to give us as normal life as was possible under the circumstances. I do not consider myself a Holocaust survivor because others had to endure much harsher conditions than we did. The worst was when after about a month, we were transferred from the Karcag ghetto to the Szolnok ghetto, where we were put in a sugar factory where they concentrated the people from all around Szolnok. It was an open-air camp; it was raining and there was nothing to sleep on at night. There was no food. People were crying. Some tried to escape and were severely beaten and we had to watch that. I was six years old and my brother was eight, so he remembers everything better than I. We were put on trains. Some trains with some of my relatives went to Auschwitz, and our train went to

---

[2] "The Nobel Prize in Chemistry 2004," *Nobelprize.org*. Nobel Media AB 2013, http://www.nobelprize.org/nobel_prizes/chemistry/laureates/2004/.

Austria. First, we were brought to a concentration camp called Strasshof, where everybody had to strip, and then we entered some chambers, which were, however, not gas chambers, because water came out of the faucets. Then trucks took us to a little village near Vienna called Guntramsdorf. There we were put into a stable with straw on the floor. Our group had about thirty or forty people. The grown-ups worked in the fields; and in the winter, they worked in a factory. The Russians arrived some time in April of 1945, and we walked back on foot to the north of Hungary. Eventually, we could take a train and arrived in Budapest, from where we went to Karcag. I remember that our house was looted, and my mother went to find our furniture, and she found a few pieces here and there. Nothing was returned voluntarily.

My father was not with us. He had been taken to the Russian front in the forced labor service. Most of their guards were tolerable, but there was a drunken major who ordered them to strip and run out into the snow. My father was captured by the Russians before the others had a chance to kill him. He returned to Hungary in 1947. He wrote in his book that there were hundreds of farmhouses in the vast land around Karcag, called *tanya* in Hungarian, and it would have been so easy to hide some Jewish families there. But there was not one case in which help was offered. My father was quite bitter about it.

What happened to your maternal grandparents?

They perished in Auschwitz. My mother never talked about them. She could never watch any movies about the Holocaust. My little cousin also perished in Auschwitz. Her mother, my aunt, survived and came back. I remembered her; before the war she had black hair and when she came back her hair was all white. Another of my cousins still lives in Szolnok, the only one of my relatives who still lives in Hungary.

I do not have very good memories of Hungary. I speak the language, but not too well. My parents decided to immigrate to Israel in 1950. My father continued to be a teacher in Israel, and became a very successful writer of math books for elementary school. He passed away in 1998 at the age of 93, and my mother passed away two years ago at the age of 91. I don't have much connection with Hungary, except for my cousin, whom I have visited twice. My brother is very different from me; although being older, he remembers more about the Hungary of the Holocaust; and he is not happy about that, but he resents Hungary less than I do, and he has a lot of friends there. He is a well-known hematologist in Jerusalem."[3]

---

[3] Ibid., 251–253.

OLEH HORNYKIEWICZ

[Dopamine success] has brought about an explosion in human brain research.

Oleh Hornykiewicz in 2003 in Vienna. (photo by I. Hargittai)

Oleh Hornykiewicz[1] (1926–) was born in Sychiw near Lviv/Lemberg (in what was then Poland and is now the Ukraine). He is Professor Emeritus of the Institute of Brain Research of the University of Vienna, Austria. Hornykiewicz showed that the lack of dopamine causes symptoms of Parkinson's disease in humans and suggested treatment with L-DOPA. In 2000, a Nobel Prize was awarded to three scientists for their discoveries concerning signal transduction in the nervous system. Hornykiewicz was not one of them, though he might have been. Two hundred fifty neuroscientists wrote an open letter protesting his omission. He has received other important awards, among them the Wolf Prize of Israel in 1979. We recorded our conversation in his office at the Institute of Brain Research in Vienna on October 25, 2003.

[1] *Candid Science V*, pp. 618–647.

Hornykiewicz started his dopamine studies on the advice of Hermann Blaschko, with whom Hornykiewicz spent time at Oxford University. Blaschko was a refugee from Germany. Upon Hornykiewicz's return to Vienna, he continued his dopamine research. The field was advancing quickly.

"By 1958, a whole group of publications came out more or less at the same time showing that dopamine was at least of some interest in the brain. Carlsson's February 28, 1958, *Science* article also presented the important observation that reserpine depleted the level of dopamine in the brain and L-DOPA restored its level; a finding also reported in May 1958 in *Nature* by Weil-Malherbe. Actually, Weil-Malherbe's study contained a more complete account than Carlsson's. Unfortunately, after publishing another full paper on brain dopamine in 1959, Weil-Malherbe did not continue these studies."[2]

As a result of his work with patients of Parkinson's disease, which he did jointly with clinicians, Hornykiewicz eventually suggested treatment by L-DOPA.

"It was easy for me to conceive the dopamine replacement idea. I remember exactly when it was. Before the paper came out, in December 1960, I was visiting Blaschko again, in Oxford, in October–November 1960, and I received the proofs of our paper sent... from Vienna. As I was correcting the proofs, the idea occurred to me, 'Why not try L-DOPA in those patients?' I had already worked with L-DOPA in Oxford when I was doing those blood pressure experiments with dopamine in guinea pigs, and L-DOPA had the same effect as dopamine. The idea of trying L-DOPA in human Parkinson patients was for me straightforward. Of course, I knew all the literature on reserpine and L-DOPA; I knew that L-DOPA also counteracted the reserpine syndrome in animals. At the beginning of 1960, there was already a paper on the anti-reserpine effect of L-DOPA in human patients. It was published by Degkwitz in Germany. He was the first to show that L-DOPA counteracted reserpine sedation, as he called it, in humans, but his patients must have had a strong akinesia, which is a prominent symptom of the reserpine-induced Parkinsonism in humans. Interestingly enough, Degkwitz, who was a neuropsychiatrist, did not think of giving L-DOPA to Parkinson patients. To me, the idea to try L-DOPA occurred immediately. Of course, I was prepared for that idea. Was it then all due to chance only? Well, wasn't it Louis Pasteur who said, Chance favors the prepared mind?"[3]

About how the idea of the L-DOPA treatment spread over the world.

"It eventually did, but it was not that simple. First, L-DOPA was a rare chemical, and it was not easy for clinicians to obtain it in sufficient amounts. Secondly, for the intravenous injection, as we used it, we usually pretreated the patients with a monoamine oxidase inhibitor, which potentiated the effects of dopamine but had its own untoward effects. This was still clinically not practicable as a treatment in a chronic condition, such as Parkinson's disease. The effect of intravenous L-DOPA in our patients was very

---

[2] Ibid., 628.
[3] Ibid., 632–633.

strong after inhibition of the monoamine oxidase, like a miracle. But it still was very short-lived, one or two hours. And we could not use higher or more frequent doses so as to prolong the effect because of the acute side effects, such as strong vomiting. So it was not until six years later that L-DOPA really became accepted and used everywhere. This happened when George Cotzias in New York had the idea and the courage to use, on a daily basis, very high oral doses of L-DOPA given in frequent intervals. That way, the high therapeutic effect could be maintained on a long-term basis. By the way, Cotzias used for his first patients DL-DOPA because it was easier to obtain in large enough quantities, and also cheaper."

Sadly, Cotzias died and his work did not continue. But the treatment of Parkinson's patients by L-DOPA has continued. However, it is still not a cure.

"It was clear from the beginning that L-DOPA was a symptomatic treatment. It is replacing dopamine, the missing substance, like insulin is used for diabetes.

Using this comparison, I would like to ask you the following question. When insulin was discovered, subsequently some people said that the discovery of insulin, which is just a treatment, in a way hindered further research on diabetes, because now there was a treatment. I wonder if you have heard of such a consideration. What would be your comment on this? Would your discovery divert attention from finding the cure for Parkinson's disease?

On the contrary. If you look at the history of research in Parkinson's disease until the discoveries about the importance of dopamine for the disorder and the replacement treatment with L-DOPA, the research was very modest. Before the use of L-DOPA, Parkinson's disease was regarded as an essentially untreatable disease. Parkinson's disease is a very severe progressive degenerative brain disorder. There were a few drugs, the so-called anticholinergics, which had a very modest effect, not more than 20 percent of improvement, and that was about all. Then the neurosurgeons developed surgical procedures in Parkinsonian patients, showing that they were to some degree effective, mostly on tremor only, and certainly not strong and persistent enough to be of use as a routine procedure. The research in Parkinson's disease was at a very low activity level, and there was no research worth mentioning on the possible causes of the disorder.

This changed dramatically after the loss of dopamine as a neurotransmitter-like substance in the Parkinson brain was discovered and it was demonstrated that the replacement of that substance with its precursor L-DOPA showed a full therapeutic effect. That was when the real research on Parkinson's disease and its possible causes actually started. When it started, it exploded, and those findings that I just mentioned stimulated research in other brain diseases. The hope emerged that similar changes could be found in other degenerative brain diseases and that treatments could be found for similar until-then untreatable diseases. Today, it is difficult for people, even for the young neurologists, to realize what the situation of the neurologists was before the L-DOPA era. The Parkinson patients were hopeless patients; they were crowding the chronic wards in the hospitals, and they ended up completely stiff and bedridden, they could not get up, they could not feed themselves, they could not move and had to be cared for until they died. The doctors were powerless. They could not do anything for those patients.

The discoveries brought about a change in all that and showed that it was possible to treat these patients and that even a chronic degenerative progressive brain disease could be treated. It has stimulated further research on new treatments and the causes of such diseases. It brought about an explosion in human brain research. This research continues at a high rate. So what has happened in Parkinson's disease research is just the opposite of what you mentioned about the possibility that the discovery of insulin may have slowed down diabetes research."[4]

---

[4] Ibid., 637–638.

FRANÇOIS JACOB

[André Lwoff] taught me that the art of research starts with finding a good boss.

François Jacob in 2000 at the Institut Pasteur in Paris. (photo by I. Hargittai)

François Jacob[1] (1920–2013), a French molecular biologist, was born in Nancy, France. World War II interrupted his medical studies in Paris. He joined the Free French Forces, participated in the Normandy invasion, and was severely wounded. He was awarded the highest French military decoration, the Croix de la Libération. He completed his medical studies after the war, but had to give up his original intention of becoming a surgeon because of his injuries. In 1950, he joined the Pasteur Institute and completed his process of becoming a scientist under André Lwoff's mentorship. In 1965, Jacob, Lwoff, and Jacques Monod were awarded jointly the Nobel Prize in Physiology or Medicine "for their discoveries concerning genetic control of enzyme and virus synthesis."[2] We recorded our conversation in 2000 in Jacob's office at the Institut Pasteur in Paris.

---

[1] *Candid Science II*, pp. 84–97.
[2] "The Nobel Prize in Physiology or Medicine 1965," *Nobelprize.org*. Nobel Media AB 2013, http://www.nobelprize.org/nobel_prizes/medicine/laureates/1965/.

This is how Jacob characterized the French school of molecular biology.

"We did mainly genetics here. I came here a few years after the war. I was heavily wounded in the war. Originally I had wanted to become a surgeon but because of my wounds in my right arm and leg it could not be. First, I quickly finished my medical school training and tried various jobs in the movies, journalism, and ended up in research, and came to the Pasteur Institute. At that time Lwoff and Monod were working in the two opposite ends of the corridor. Lwoff was working in the physiology of bacteria and Monod was doing some mixture of physiology and biochemistry. We started doing molecular biology as a derivative of physiology, focusing on the properties of the bacterial cell. Lwoff was working on bacteriophage, Monod on the induction of enzyme synthesis, and with Wollman we started on the genetics of bacteria. We had a system in which we studied the movement of the chromosomes between what we could call male and female, or donor and acceptor. The male chromosome is injected into the female and it moves at a constant speed, and it is possible to map the movement of the chromosome by the time of its entry. It is like when a train goes from Paris to Marseille at a constant speed. You can plot the position of the train at any given moment if you know the starting time and the speed. We had several different ways of looking at the introduction of a chromosome into bacteria, we could do genetic recombination or we could do mapping by time.

Ours was a fantastic system for playing with physiological genetics. Generally, when you do genetics, you have a male and female with markers, and you look at these markers after recombination, in the progeny. We had a new approach in that we could look at the gene expression before recombination. We could show, for example, that when we had what we called the prophage, a form of the bacteriophage genome in which the DNA of the phage is integrated into the chromosome of the bacterial host of the phage. As soon as it entered the cytoplasm of the recipient cell, it started multiplying. This multiplication started without genetic recombination. We could determine whether we had the gene controlling the synthesis of an enzyme, we could measure the time required to start the production of the enzyme once the gene was in the female. We could do rather precise kinetics on that, and we could do the same thing with the regulatory gene. We had a system worked out by Monod: there was an enzyme, β-galactosidase, which was manufactured by the bacteria only in the presence of lactose of β-galactosides. Without it, the medium just did not synthesize the enzyme. We could combine a male, which had a gene which managed to prevent the enzyme from being synthesized in the absence of lactose, with a female, which did not have this gene, and see how it was expressed. In other words, we had a system which allowed analysis of various bacterial functions. It was rather different from going straight to the gene-making sequences of DNA or the peptides."[3]

Commenting on science in France.

"France is no longer such an important power in science as it used to be, and there are several reasons for that. The structure of the French scientific system is not very

---

[3] *Candid Science II*, 88–89.

favorable for scientific discovery. It is an old system, which has been progressively transformed since the war. We used to have the great professors to whom the students were afraid to talk. My first visit to the United States was in 1953, and I was extremely surprised by their relationship between professors and students. It was very relaxed; whereas it was very rigid in the European countries, perhaps less so in England."[4]

[4] Ibid., 90.

AARON KLUG

Research is not just going from mountain top to mountain top; you also have to work in the valleys, and that takes time and freedom.

Aaron Klug in 2000 with his virus model at the MRC Laboratory of Molecular Biology in Cambridge, UK. (photo by I. Hargittai)

Aaron Klug[1] (1926–), British biophysicist and molecular biologist, was born between the two world wars in Lithuania, when this small Baltic state was independent. His family soon moved to South Africa. Klug received his bachelor of science degree from the University of Witwatersrand in 1945, his master of science degree from the University of Cape Town in 1946, and his PhD degree from the University of Cambridge in the UK in 1952. He was at Birkbeck College, University of London, between 1954 and 1961 and has been

---

[1] I. Hargittai and M. Hargittai, *Candid Science II*, 306–37.

with the MRC Laboratory of Molecular Biology (LMB) since 1962. In 1982, Klug was awarded the Nobel Prize in Chemistry "for his development of crystallographic electron microscopy and his structural elucidation of biologically important nucleic acid-protein complexes."[2] In addition to research, Klug was active in science administration and served as President of the Royal Society (London) between 1995 and 2000. We recorded a conversation in 1998 and refined its transcripts during several sessions in 2000 in Cambridge.

Structure determination and methodological innovation seemed to have had equal weight in Klug's research.

"Absolutely. Without the new methodology, we would not have been able to solve the structures. When we started out to use electron microscopy for our virus structure studies (to complement the X-ray work), we thought that we would just get an overall view of the structure. People had not realized that you could extract much more information from electron microscope images. In 1969, when we demonstrated the first three-dimensional image reconstruction of virus particles and bacterial flagella at a meeting in New York, Sturkey, an electron diffractionist, called it a 'load of crap.' What he did not understand was that we were dealing with relatively weakly scattering objects. We had demonstrated this, for example, on multilayers of cell walls, which could be analyzed by electron microscopy and diffraction, taking into account the overlapping layers. A tremendous amount of background work went into three-dimensional image reconstruction before we began to understand what was going on. Sturkey had a theoretical point that we were ignoring multiple scattering, but we justified it. Because of your own research in electron diffraction, you will understand the point about multiple scattering, and so do I, and so did Sturkey. I have a PhD in physics and do know some physics and mathematics.

When you say 'methodology,' I started out with electron microscopy with the simple view that it was well understood. In fact, I realized soon enough that this was not the case. I had the good fortune to be a nonexpert. When you are a nonexpert, you do not come with many presuppositions. There were various people working in the field trying to get 'the perfect picture,' and I realized that there was no such thing. I introduced the approach of taking a series of micrographs in the electron microscope at various degrees of defocusing and then correcting them for the contrast transfer function. With this method, we could create an image of transparent objects, that is, nonstained biological specimens.

The technology developed in the course of practical studies. We started out with a real problem in a helical virus, and I soon realized that we could make a three-dimensional reconstruction by using the theory of helical diffraction. Every time you tilted the specimen, you could recover another function in the mathematical expansion of the electron density. I had developed the approach for Rosalind Franklin's X-ray studies of the tobacco mosaic virus [TMV]. Later, I saw that it was a special case of a more general principle in Fourier theory.

---

[2] "The Nobel Prize in Chemistry 1982," *Nobelprize.org*. Nobel Media AB 2013, http://www.nobelprize.org/nobel_prizes/chemistry/laureates/1982/.

The method also became the basis of the principle of the X-ray CAT scanner. Hounsfield and Cormack received the Nobel Prize in 1979 for computer-assisted tomography [CAT]. Hounsfield read my *Nature* paper of January 1968 and took out a patent for CAT in August 1968. I had realized earlier that you could apply it to medical radiography, and I went to see some radiographers, but they told me that they did not need this 'fancy stuff.' They said, 'We understand exactly what we see in a medical X-radiograph,' but I asked them about the radiation doses used.

The X-ray tomography of the time used a moving source of X-rays and a moving film, so all the density, except in one purely geometrical optics plane, was out of focus, the rest blurring the radiograph. I said to them: 'Look, you are giving a much bigger dose than you have to, and the question is how much more information do you get?' Later, it was shown mathematically that you could get more information for a given dose by image reconstruction by CAT scanning than by X-ray tomography, and CAT scanning has become the standard method. Some people think that I should have got the Nobel Prize with Hounsfield. This story illustrates the important point in science that you sometimes find the solution to a problem from another field."[3]

There was a slight controversy in preparing Klug's Nobel lecture for publication.

"A. N. Whitehead was a famous philosopher who coauthored *Principia Mathematica* with Bertrand Russell, and he said, 'It is more important that an idea be fruitful than that it be correct.' When I put together my Nobel lecture for publication, the editor wanted to cut out the picture depicting our initial idea of nucleation. He said it was wrong. I replied that it was indeed wrong in detail but everything essential was in there, so including it would show how science is a *process* of establishing the truth. The protein disk is an obligatory component for the formation of the virus. It performs two simultaneous functions. One is starting the physical assembly of the protein subunit, nucleation. At the same time, it recognizes a special sequence of the viral RNA, determining the specificity of interaction.

Many people think that science is just the application of various formulas. Some of it is, but they need to understand that, in a developing field, there are many steps on the way, and you can sometimes take the right step for the wrong reason. I worked on TMV from 1954, when I joined Rosalind Franklin, to the 1970s, when we were finally able to prove the mechanism of the assembly. Even after many years, this is the most detailed system of its kind that has been worked out. This was an important achievement, and for me, TMV was my first major scientific adventure."[4]

On what the next frontier of science will be.

"That is fairly obvious, although I do not know how long it will take, but neuroscience, the workings of the nervous system, and particularly the brain. If I were starting over

---

[3] *Candid Science II*, 308–11.
[4] Ibid., 313.

again, I probably would go into neurobiology. Whether I would have the taste for those kind of experiments, I do not know, but I find neural networks fascinating.

My scientific life has been one in which I worked on relatively messy systems, which physicists would not touch. On the other hand, I was able to bring some rigor into them by doing things properly and by developing new techniques, and it suited me very well. Those of us here in the Lab [MRC LMB] have helped create the subject, so it bears our image. But now cell biology has become so important, and I do not know whether you can make any progress with the brain without finding out more about the interactions between sets of molecules. One just has to start.

Thinking about science, you have to start even though you do not know where the end will be. I have been lucky because I had a good preparation without planning it and started my career when a new subject was opening up here, and have had the opportunity to work for the MRC. This is a very enlightened body, which has let me work on long-term projects. Research is not just going from mountain top to mountain top, you also have to work in the valleys, and that takes time and freedom."[5]

[5] Ibid., 328–29.

ARTHUR KORNBERG

My respect for science is unqualified.

Arthur Kornberg in 1999 at Stanford University. (photo by I. Hargittai)

Arthur Kornberg[1] (1918–2007) was born in Brooklyn, New York, graduated from the City College of New York in 1937, and received his doctor of medicine degree at the University of Rochester in 1941. He was affiliated with the National Institutes of Health, (1942–1953) and Washington University School of Medicine in St. Louis, Missouri, (1953–1959) before joining Stanford University, where he spent the rest of his career. Arthur Kornberg and his first wife and fellow scientist Sylvy Kornberg (née Levy) had three sons. Kornberg was awarded the 1959 Nobel Prize in Physiology or Medicine jointly with Severo Ochoa "for their discovery of the mechanisms in the biological synthesis of ribonucleic acid and deoxyribonucleic acid."[2] We recorded our conversation in 1999 in his office at Stanford University.

---

[1] I. Hargittai and M. Hargittai, *Candid Science II*, 50–71.
[2] "The Nobel Prize in Physiology or Medicine 1959," *Nobelprize.org*. Nobel Media AB 2013, http://www.nobelprize.org/nobel_prizes/medicine/laureates/1959/.

Istvan Hargittai, Arthur Kornberg, and James D. Watson in 2001 in Stockholm.
(photo by M. Hargittai)

One of the Kornbergs' three sons, Roger Kornberg received the 2006 Nobel Prize in Chemistry "for his studies of the molecular basis of eukaryotic transcription." In 1999, we asked Arthur Kornberg about his influence on his sons.

"Hard to tell. My eldest son Roger, now a distinguished scientist, was fascinated with science from an early age. When we asked him at nine, 'What do you want for Christmas, Roger?' his response was 'a week in the lab.'[3]

My second son, Tom, never entered the lab and studied the cello full time. He attended Juilliard and was a student of Leonard Rose. His classmates and still close friends include Yo-yo Ma and Emanuel Ax. Tragically, Tom developed neuromas on his left index finger and had to give up playing. While at Juilliard, he was also a full-time student at Columbia College. He took chemistry, physics, and biology. While there in 1969–1970, John Cairns found a mutant of *E. coli* that appeared to lack DNA polymerase but could still make DNA. My work came into question in the minds of many people. How could replication go on in the absence of this enzyme. *Nature* magazine led a chorus that DNA polymerase was a 'red herring.' Tom proved to be a gifted experimentalist. Within a few months, he discovered novel DNA polymerases. The one that I'd discovered first had a role in replication, but it was largely involved in repair of DNA. All three DNA polymerases have the same mechanism but are used for different things in a growing cell. After these electrifying discoveries, after only a few months in the laboratory, Tom went on to graduate school and later to postdoctoral work in molecular genetics. He is now professor at the University of California in San Francisco, doing very important work in developmental biology.

---

[3] Roger Kornberg won the 2006 Nobel Prize in Chemistry for his discoveries about how genetic information from DNA is copied to RNA.

My third son, Ken, is an architect. But at Stanford he took physics, biology, chemistry as well as architecture and mechanical engineering. Now he is the most gifted designer of laboratories and very much sought after. He has a profound understanding of the science that his brothers and I do and a flair for design.

Did I influence them? My wife and I influenced them by our love for and devotion to science. Naturally, you influence your children in many ways. But I never discouraged Tom from being a cellist. I admired and applauded his career in music. I've been equally supportive of Ken as an engineer and architect. I've a profound admiration for what they're doing. From the very earliest age, when they were still in diapers, I would take them on trips with me. Later on, there were more extensive trips, week-long trips when I was visiting professor somewhere. Even though I was busy and consumed by my preoccupation with time, my sons will tell you that I always had time to be with them. These trips were great, for them and me. With three boys, so close in age and highly competitive, it was good to take one to be the center of my and other peoples' attention. Now they're very close friends, devoted to each other. There's nothing more that a parent can wish for."[4]

---

[4] *Candid Science II*, 64–65.

PAUL C. LAUTERBUR

The experts are very proud of their expertise and disappointed when something proves to be original that they had not thought of themselves and they like to look for flaws....

Paul C. Lauterbur with his wife, Dr. M. Joan Dawson, in 2005 in Lindau, Germany.
(photo by I. Hargittai)

Paul C. Lauterbur[1] (1933–2007) was born in Sidney, Ohio. For many years, he directed the Biomedical Magnetic Resonance Laboratory at the University of Illinois in Urbana. He and Peter Mansfield of the University of Nottingham shared the 2003 Nobel Prize in Physiology or Medicine "for their discoveries

---

[1] B. Hargittai and I. Hargittai, *Candid Science V*, 454–79.

concerning magnetic resonance imaging."[2] Lauterbur received his bachelor of science degree in chemistry from the Case Institute of Technology in Cleveland, Ohio. Then, he decided that it was not important to continue his formal studies and concentrated on research. In 1962, he obtained his PhD degree from the University of Pittsburgh. Between 1969 and 1985, he was Professor of Chemistry at the State University of New York at Stony Brook. He moved to the University of Illinois in 1985. The Nobel Prize was slow to come to him. In a 2003 publication, the Nobel laureate NMR spectroscopist Richard Ernst noted: "The disrespect for MRI in Stockholm is particularly difficult to understand."[3] We recorded our conversation in 2004 in the Lauterburs' home in Urbana, Illinois.

It was not easy for Lauterbur to have his initial publication about NMR imaging accepted.

"This was a form of imaging that was different from anything that had been done before. That's also why the first version of my paper to Nature was rejected. It was because I left out cancer and medical diagnoses and concentrated only on physics, on science, basically. That was also why I gave it a strange name, NMR Zeugmatography.

Did Nature *give you an explanation for the rejection?*

I have the letter somewhere, but I have not looked at it for some time. It was just the standard text, something to the effect that we don't see why you make such a big fuss about it.

But it was published nevertheless.

It was.[4]

How did you convince them to publish it?

I wrote them a long, impassioned letter and offered to put in more speculation about possible applications to enforce the idea that there may be some interest in it. Someone else later told me that my manuscript was then sent out to a new referee whose comments were not terribly encouraging but then he added something to the effect that this seems to be crazy but I had never done anything crazy before.

That means that by then you had established your reputation.

I had been very active in nuclear magnetic resonance for years, ever since I was a graduate student. I was forty-three years old when I submitted the manuscript to *Nature*.[5]…

It seems to be easier to publish mediocre papers than groundbreaking papers.

---

[2] "The Nobel Prize in Physiology or Medicine 2003," *Nobelprize.org*. Nobel Media AB 2013, http://www.nobelprize.org/nobel_prizes/medicine/laureates/2003/.
[3] Richard R. Ernst, "Foreword." In H.-J. Müller, P. K. Madhu, and R. R. Ernst, Eds., *Current Developments in Solid State NMR Spectroscopy* (Wien and New York: Springer, 2003).
[4] P. C. Lauterbur, "Image Formation by Induced Local Interactions: Examples Employing Nuclear Magnetic Resonance." *Nature* 242 (1973): 190–91
[5] *Candid Science V*, 456–57.

It is very natural because for a truly original paper, there are no real peers to judge it. The experts are very proud of their expertise and disappointed when something proves to be original that they had not thought of themselves, and they like to look for flaws and tend to think that it may not be true. There may be all sorts of psychological reasons. When I submitted a proposal to NIH, they examined it at a study session— the usual way they review proposals. Although people are not supposed to talk about it, someone told me that at first the reviewers were negative because they found my proposal crazy. Then someone said that maybe because it is crazy they should take a second look, and, fortunately, they did. The consensus was that they still found it crazy, but they could not find anything wrong with it, so they decided to fund the proposal."[6]

---

[6] Ibid., 474.

JOSHUA LEDERBERG

Our brain has enough plasticity to enable a good society.

Joshua Lederberg in 1999 at Rockefeller University. (photo by I. Hargittai)

Joshua Lederberg[1] (1925–2008) was born in Montclair, New Jersey, and grew up in New York City. He graduated from Columbia College in 1944 and received his PhD degree from Yale University in 1947. Following periods at the University of Wisconsin and Stanford University, he served as President of Rockefeller University between 1978 and 1990, after which he stayed on as researcher. In 1958, Lederberg received half of the Nobel Prize in Physiology or Medicine "for his discoveries concerning genetic recombination and the organization of the genetic material of bacteria."[2] We recorded our conversation in 1999 in his office at Rockefeller University.

---

[1] I. Hargittai and M. Hargittai, *Candid Science II*, 32–49.
[2] "The Nobel Prize in Physiology or Medicine 1958," *Nobelprize.org*. Nobel Media AB 2013, http://www.nobelprize.org/nobel_prizes/medicine/laureates/1958/.

Lederberg was one of two scientists (the other was Erwin Chargaff) who immediately saw the significance of Avery and colleagues' report about DNA being the genetic material, and he built his research on it. Here is how Lederberg reflected on its importance.

"It had an enormous influence on me, which I have documented. But to answer your first question first, I had probably heard about it even before it was published. Although Alfred Mirsky has been criticized for being so negative and being so skeptical, at the same time he was the herald, he was the person who was carrying the news of it to many, many people. He was pointing to the work and said, 'Look, these people are doing this work; they make these claims that DNA alone is sufficient for the transforming activity. I, Mirsky, am skeptical about that. It's very difficult to be sure that there isn't some residual quantity of protein and that the protein is the active factor.' It was a perfectly legitimate position.

I don't think he should be criticized for maintaining that skepticism. He maintained it for a long time, and we may say too long, but it was important to maintain some degree of resistance in order to evoke a response. The response was: Mac McCarty did a wonderful job in piling up one bit of evidence after another that really nailed it down, for sure, about the purity of the material. You could not be absolutely sure until DNA was synthesized. As long as you were preparing DNA from a natural source, you could always argue that there was also some molecule of something else that was contaminating the DNA and was the source of specificity. There was no way around that.

When Khorana and Kornberg synthesized DNA, we could be really quite certain that it was just DNA and DNA alone that was sufficient. In 1952, there was the Hershey-Chase experiment, and in 1953, there was the X-ray structure of DNA. Watson and Crick didn't add any more evidence as to whether it was DNA but they made it plausible, the physical model of how replication could take place. By about 1955, there was nobody arguing with it.

In 1944, I was a student at Columbia. Mirsky was collaborating with one of the professors at Columbia (A. W. Pollister) and there were frequent communications, many seminars; so I knew about the paper by the time it was published. We had a reprint of the article, though not the journal. The *Journal of Experimental Medicine* was up at the medical school. I also have a written record because I was so impressed by the article that I wrote a note about it on January 20, 1945,... that I have saved, and it's on the website [http://profiles.NLM.nih.gov]. There are not many contemporary records which show that people wrote down that that was their view, but I've been in a long-standing argument with Gunther Stent about the reception of the Avery story. That website, again, documents every element of that argument.

I was not the only one who was influenced. In terms of who changed research directions, what was there to do? There were two questions. One, was this pure DNA? Or was this some contaminant? The assay system was quite arcane, and very few people who were not already experts on *pneumococcus* would have been willing to try to repeat those experiments themselves. First of all, it's a dangerous organism to work with. Second, the recipes that were published for how to deal with it were very complex in the early days. I would have been very reluctant to try to attempt a direct repetition unless I had the opportunity to actually work for a while in the Avery laboratory. That lab closed down within a couple of years after their report, Avery died in 1955, and that was another reason that slowed down the diffusion of the experiment per se.

The other question was, here is a phenomenon but is it really the gene? It's a very specialized story, it had to do with one trait, the capsule of polysaccharide. For a long time, there were many people willing to believe that this was a phenomenon of seeding, of having a template for the aggregation. There were other examples of it. If you use starch phosphorylase with glucose 1-phosphate, you need a little starch as a starter in order to get the assembly, in order to get the polymerization into starch. So other polymer syntheses were known that needed polymer starters. One could have argued that that's what this phenomenon was, maybe it needed some nucleic acid as a kind of a coenzyme, so other interpretations were possible in addition to the one that this really was the gene. That was the question I wanted to address to really nail this down because I did see that here was the beginning of what was by then called genetic chemistry and everybody now calls molecular genetics. We needed to solidify the two pillars: one, is it DNA? What's the chemical identity? And, two, what's the phenomenon in the biological sense? Is it really gene transfer? Neither of those was totally solidified by the observation of this one case, of this one organism, of this one trait, in the *pneumococcus*.

Chargaff went after the chemistry of the DNA, and it was a very important contribution to break down the dogma of the 1:1:1:1 relationship of the tetranucleotide structural hypothesis that Levene had developed here at the Rockefeller Institute. He had made many important contributions but the best thing he was able to do about the structure was to conclude that it was either a tetranucleotide or a tetranucleotide polymer because of the approximate uniformity of the ratios of the bases that were present in his DNA preparations. When Chargaff demonstrated that DNA was not 1:1:1:1, we began to see that there was more room for variety of internal structure. That made much more plausible the possibility that nucleic acids could be the carrier of genetic information.

On the other side, if this was a gene, my first thought was that we needed to get a transformation in an organism whose genetics had already been worked out. I thought *Neurospora* would be a very good candidate for that. I had already been working with Ryan on mutants of *Neurospora*.

Ryan had been a postdoctoral fellow with Beadle and Tatum in 1941/2 at Stanford. He was their first postdoctoral fellow after their initial publication, in the fall of 1941, on biochemical mutants in *Neurospora*. Ryan learned the technology and the ideology of biochemical mutations at Stanford and brought that back with him to Columbia. I met him when I was a sophomore in the academic year 1942/3. I'd heard about him from other people around the Department and it struck me that this is the work I really wanted to do. I gave him no option but to accept me in his laboratory. I was a real pest."[3]

---

[3] *Candid Science II*, 37–39.

RITA LEVI-MONTALCINI

The intellect of man is forced to choose, Perfection of life or of the work.
—William Butler Yeats, "The Choice"; epigraph in
Levi-Montalcini's *In Praise of Imperfection*.

Rita Levi-Montalcini in 2000 in Rome. (photo by M. Hargittai)

Rita Levi-Montalcini[1] (1909–2012) was born in Turin, Italy. She was Director Emeritus of the Institute of Cell Biology of the Italian National Council of Research (CNR) in Rome. Levi-Montalcini shared the 1986 Nobel Prize in Physiology or Medicine with Stanley Cohen of Vanderbilt University School of Medicine, Nashville, Tennessee, "for their discoveries of growth factors."[2] Nerve growth factor (NGF) is not only necessary for the survival of certain nerves but also regulates the directional growth of the nerve fibers.

Rita-Montalcini came from a professorial family, but her father was reluctant to support her going to medical school. Finally, he relented, and she

---

[1] I. Hargittai and M. Hargittai, *Candid Science II*, 365–75.
[2] "The Nobel Prize in Physiology or Medicine 1986," *Nobelprize.org*. Nobel Media AB 2013, http://www.nobelprize.org/nobel_prizes/medicine/laureates/1986/.

Rita Levi-Montalcini with Viktor Hamburger. (courtesy of R. Levi-Montalcini)

graduated as a doctor of medicine from the University of Turin. However, the anti-Jewish laws of fascist Italy prevented her from practicing medicine. Shortly after receiving her Nobel Prize, Rita Levi-Montalcini wrote a beautiful autobiography, *In Praise of Imperfection*.[3] We recorded our conversation in 2000 in Levi-Montalcini's office in Rome.

We asked Levi-Montalcini about her books. Elogio dell'imperfezione (*In Praise of Imperfection*) *has been translated into English.*

"Unfortunately, all my other books are in Italian, neither of them was translated into English, but some Spanish and French translations have appeared. The last one, to come out soon, is a selection of letters I wrote to my mother about my work at the time I made the discovery—that is half a century ago, when I lived in the United States. It is a day-by-day sequence of letters. Its title is *Cantico di una vita,* something like 'the Hymn of Life.' The letters show that I was full of enthusiasm. We have just recently found about fifteen hundred letters in a wooden crate in the basement of our house—I thought they were lost. I selected about 200 of them, and edited them into this book, perhaps the best book I have ever written. The letters are all long and neat letters, some of them are eight pages long, and not one word is crossed out or changed in them.

---

[3] Rita Levi-Montalcini, *In Praise of Imperfection: My Life and Work*, trans. Luigi Attardi (New York: Basic Books, 1988).

Now I am working on another book. There is a beautiful, almost biblical poem by Yeats, 'The Second Coming,' but it is very pessimistic about the future. I am not like him; I have a basically optimistic attitude, so the title of the book I am writing now is *The New Coming*. It is about the possibility of saving the human species (and other species) by changing the relationship to children, to women, and to young people. This is the only hope to survive.

Perhaps the most interesting book I have written, and it is in English, *The Saga of the Nerve Growth Factor*, it is a collection of all the papers and articles that have appeared on this topic and some of which are no longer available."

About her road to science.

"Oh, I believe I have never become a scientist. I am more of an artist. I just use the nervous system as the area of my study. My approach to science is this: you can do it! I was a student of Guiseppe Levi, an outstanding Italian scientist. Three Italians have gotten the Nobel Prize, Luria in 1968, Dulbecco in 1975, and myself in 1986. All three of us have become scientists just because we were students of Guiseppe Levi. He himself never received the Nobel Prize.

But I don't believe that I have ever been a scientist. My twin sister is an artist, a painter, one of the best in Italy. My brother was an architect and an excellent sculptor. I believe that my approach to science was from the point of view of the beauty of the nervous system and not just plainly because I was interested. Still now, I don't believe I am a scientist, I approach science more from an artistic point of view than from a scientific one. Ever since I entered into the field of science as a young person, I have never thought that I would become a scientist. I was more interested in social problems than in science. It was just my good luck that I was prevented from continuing my work as a medical doctor. Thus I had to work in my bedroom in a little laboratory, and that is where I did the first steps of the work that later led me to Stockholm.

My interest in science was just casual and I owe it to Benito Mussolini who prevented me to become a regular doctor, because I was not Aryan.[4]..

...I came form a very cultured, very highly educated Jewish family. We were absolutely not observant, my father and mother never went to the synagogue, they did not care about that. I only became Jewish when the law came against us, before I never knew I was one. I had both Jewish and Catholic friends and it never was a problem. Even after the racial laws, they remained friends. People did not much care about these laws, of course, later many became anti-Semitic because they wanted to make success in life. Now I feel Jewish. But I became Jewish only when the racial laws came, after the persecution, not before.

Did you know Primo Levi?

He was a splendid person. He wrote the best book that will ever be written. I am sure, absolutely sure, that he never committed suicide. He may have lost his balance because he was very weak, and perhaps that is why he fell.... The family insisted that it

---

[4] *Candid Science II*, 366–68.

was suicide, but I am sure it was not; I knew him too well. He should have been interviewed about that time; and he was absolutely far away from even remotely considering of taking his own life.

We were good friends but we did not meet often, he lived in Turin and I was in Rome. I talked with him shortly before he died, I told him, Primo, come to Rome; we could go to the same psychiatrist. It was just after my Nobel Prize and I was upset by too much popularity, which I could not handle well. So I went to see a psychiatrist. But Primo said no, I would never leave my mother, not even for one hour—and then he fell down the stairs. It is absolutely out of the question that he killed himself."[5]

[5] Ibid., 370.

EDWARD B. LEWIS

[Genetics] is a twentieth-century subject whose importance for the study of biology has been slow to be recognized.

Edward B. Lewis in 1997 at Caltech. (photo by I. Hargittai)

Edward B. Lewis[1] (1918–2004) was born in Wilkes-Barre, Pennsylvania. He received a bachelor of arts degree in biostatistics at the University of Minnesota in 1939, and a PhD degree in genetics in 1942 and master of science degree in meteorology in 1943, both at the California Institute of Technology. He spent his entire career at Caltech in Pasadena. Lewis shared the 1995 Nobel Prize in Physiology or Medicine with Christiane Nüsslein-Volhard and Eric

---

[1] I. Hargittai and M. Hargittai, *Candid Science II*, 350–63.

F. Wieschaus "for their discoveries concerning the genetic control of early embryonic development."[2] We recorded our conversation in 1997 on the campus of Caltech.

We asked him about genetic engineering.

"It's wonderful.
  It could be misused.
  It could be. But there are bound to be great benefits of genetic engineering, especially in diagnosing diseases and developing cures. Vigilance will be needed to be sure that genetic engineering is not abused.
  Not only have scientific discoveries been abused in the past, scientists themselves have often been persecuted for their beliefs. A tragic chapter in recent times was the persecution of geneticists in the Soviet Union during the Stalin era by Lysenko and his followers. Lysenko claimed that heredity was not controlled by genes and chromosomes but by the environment. He gained power because such a view supported Marxism-Leninism, which denied that there could be a genetic as well as an environmental component to human behavior. Lysenko's claims could not be substantiated, but meanwhile, geneticists in the Soviet Union were persecuted, ridiculed as 'Morgan-Mendelists,' and prevented from teaching genetics or doing research in it. Such persecution of a science is a warning to the whole scientific community that it can happen even in modern times. I find it alarming that there seems to be a movement on to attack the whole idea of genetic engineering.
  Can genetic engineering make a better human being?
  This question is too loaded with ethical and other implications that I would rather not discuss.
  Could genetics data about intelligence be useful?
  Yes, but we need much more information about the genes involved. Anything as complex as intelligence will be determined by a multitude of genes interacting with one another. It is well-known that persons who have an extra twenty-first chromosome have Down's disease and are quite handicapped both mentally and physically. What does this tell us about genes involved in intelligence? It tells us a lot. First, the twenty-first chromosome, although very small, still has many hundreds of genes. Three sets of these genes, even though none of the genes need be defective, is sufficient to disrupt virtually every organ system of the body, and the brain is no exception. Fortunately, for most of the larger chromosomes the effects of an extra chromosome are so drastic that the individual dies before birth. In another example, persons who have an abnormal number of sex chromosomes are at risk of mental problems. Females have two X chromosomes and males an X and a Y chromosome. Persons who are XXY are not only intersexual, they tend to be of lower intelligence. Males with two Y

---

[2] "The Nobel Prize in Physiology or Medicine 1995," *Nobelprize.org*. Nobel Media AB 2013, http://www.nobelprize.org/nobel_prizes/medicine/laureates/1995/.

chromosomes instead of one also are at higher risk of having mental problems, including in some cases... overly aggressive behavior.

Single gene defects have been identified in flies that have effects on the fly's behavior, or memory, or ability to learn. Using DNA from such genes, the homologous genes have in some cases been already identified in human beings. This is a start toward learning what such genes do in people.

Pedigree and twin studies of intelligence in people have shown that many genes must be involved in producing the wide range of intelligence exhibited in the general population that are genetically based—of course, cultural or other environmental influences play a big factor, too, but are not exclusively involved by any means.

If you mean can we expect to use genetics to improve intelligence, that is really science fiction. Literally thousands of genes are involved in determining how the brain functions. Above- or below-average intelligence, then, follows the same rules that govern other complex traits, such as human stature or yields of corn, for example. It is the aggregate of many genes, each with a relatively tiny effect on brain function that determines whether a person's genetic endowment will confer above- or below-normal intelligence."[3]

---

[3] *Candid Science II*, 361–63.

PETER MANSFIELD

I used to spend hours at night half asleep and half dreaming about solving particular problems.

Peter Mansfield in 2005 at the entrance to the Sir Peter Mansfield Magnetic Resonance Centre at the University of Nottingham, UK. (photo by I. Hargittai)

Peter Mansfield[1] (1933–) was born in London and is Emeritus Professor of Physics at the Sir Peter Mansfield Magnetic Resonance Centre, University of Nottingham. He was co-recipient with Paul Lauterbur of the 2003 Nobel Prize in Physiology or Medicine "for their discoveries concerning magnetic resonance imaging."[2]

Initially, the rigid British system of education of his time failed Mansfield. He had attended a so-called ordinary school, which he completed at the age

---

[1] I. Hargittai and M. Hargittai, *Candid Science VI*, 216–37.
[2] "The Nobel Prize in Physiology or Medicine 2003," *Nobelprize.org*. Nobel Media AB 2013, http://www.nobelprize.org/nobel_prizes/medicine/laureates/2003/.

of fifteen, but from which he could not go on to higher education (this system since has been changed). Mansfield was determined to become a scientist and for years he worked and studied simultaneously to catch up with the educational system. He succeeded, but he was twenty-six years old when he received his bachelor of science degree in 1959. From this point, things progressed in a normal way and by 1962, he had obtained his PhD degree. Both his degrees were from Queen Mary College, University of London. He has been with the University of Nottingham since 1964, where he became professor in 1979, and from which he took early retirement in 1994. We recorded our conversation in 2005 at the University of Nottingham.

MRI, magnetic resonance imaging, was originally known as NMRI, nuclear magnetic resonance imaging. We asked Mansfield how the N got dropped from the name

"That was due to the Americans; it was dropped probably in the mid-1980s. There are a couple of explanations for why it was dropped. The official reason was that there was squabbling in America with the nuclear medicine people. Because NMR had the word nuclear in it, they wanted it to be in their department. The radiologists in the States, which have an even more powerful lobby, did not want the word nuclear in the title because they wanted imaging to be in radiology. That's what then happened. The Americans decided between themselves that there should not be NMR imaging, it should be MR imaging. We played no role in it, although much of the development happened here in Nottingham, and not in America.... Then, in the early 1980s, it started taking off in America. Like most things, if the Americans get involved, it is great—and very soon it will be considered to have been invented in America. They did it first, never mind what happened in Europe. That's what happened with MRI.

It happened with the manufacturing as well, because companies got interested in the 1980s. One of the first...was a subsidiary of Johnson & Johnson, which was involved for five years before it decided to pull out. In Europe we had Siemens, Phillips, EMI, and a whole range of smaller companies. The very last company to come into the MRI business was General Electric. They came in in the mid-1980s; they had waited and waited before they came in. When they did come in, they dominated completely, and they are now making well over half the total number of MRI systems in the world. They got the biggest share in the market. It should have been Europe, but that was not to be.[3]...

In your two-minute Nobel banquet speech you mentioned that you receive many letters of gratitude from patients, but also that a few mention the claustrophobic effects of the tight confinement in the machine, and some mention the rather high level of acoustic noise during the scanning process. Then you added that these problems are not being ignored. Are you involved in looking for ways to ease this situation?

My small company is involved in research to reduce the level of the acoustic noise in the MRI machines. Have you ever been inside such a machine? They are very noisy.

---

[3] *Candid Science VI,* 224–25.

Every time the gradient switches on and off, considerable noise is generated. First, we are concerned with the question, why is that noise associated with the gradient change? Secondly, can we do anything to reduce it? Then another thing that the company is particularly interested in is the reduction of the electric field, which is automatically created as described by Maxwell's laws of electromagnetism. An electric field is associated with the magnetic field. When that electric field is created, it is strong enough to cause currents to flow in the body. It can cause muscular twitch, so every time we switch the gradient, the body twitches. That is not good for patients. There are ways of reducing the electric field associated with the magnetic-field gradient, by not switching the gradient too fast. If the gradient is switched on and off more slowly, the electric-field component can be reduced. The problem is if the magnetic field gradient is switched more slowly, then the time taken to do the imaging is increased. Therefore that is not the approach we wish to take. My company is looking at ways of reducing this electric field. They are the two major problems we are working on, and there may be others as time goes on."[4]

Peter Mansfield took early retirement.

"I took early retirement because I wanted to spend more time on research. I've done this, but I think I've spent too much time on this particular area that I mentioned, reduction of acoustic noise. It's an important problem and it would be nice to solve it. In my case what's happened is that I have made some progress, but I have not solved the problem completely. The question is how much time should one spend on a problem before you give up? You either solve it or you don't solve it. Unfortunately, my character is such that I find it very difficult to give up on a problem. I keep thinking, maybe tomorrow I'll solve it.

Is this stubbornness or perseverance?

I'm stubborn, that's it; that's the fault. If you're too stubborn, you end up wasting time, and that's what I've done. I've identified another problem, which is the E-field problem, but the acoustic problem still remains to be solved. We have some ideas how one might do it; but I've spent fifteen years on this problem. That's almost more time than I'd spent on the whole of NMR and MRI together. The time it took to come up with the original ideas for MRI took me four or five years. It wasn't one idea either; it was a whole world of ideas. At the end of my life I've spent fifteen years on one problem and I still haven't solved it."[5]

---

[4] Ibid., 229–30.
[5] Ibid., 236.

MACLYN MCCARTY

I consider the Watson-Crick paper one of the major steps after our discovery.

Maclyn McCarty in 1997 in his office at Rockefeller University. (photo by I. Hargittai)

Maclyn McCarty[1] (1911–2005) graduated from Stanford University in 1933 and earned his doctor of medicine degree from Johns Hopkins University in 1937. He spent his professional career at the former Rockefeller Institute, now Rockefeller University, in New York City. When we met, he was Professor Emeritus. He worked with Oswald T. Avery in the early 1940s on what then was called the "transforming principle." It turned out to be DNA—this was their principal finding. Colin M. MacLeod was also a member of the team.

Their seminal paper showed for the first time that DNA is the genetic material.[2] Following his years with Avery, McCarty spent a successful research career investigating proteins, the biology and immunochemistry of streptococci, and

---

[1] I. Hargittai and M. Hargittai, *Candid Science II*, 16–31.
[2] O. T. Avery, C. M. MacLeod, and M. McCarty, "Studies of the Chemical Nature of the Substance Inducing Transformation of Pneumococcal Types," *Journal of Experimental Medicine* 79 (1944): 137–58.

rheumatic fever. He was member of the National Academy of Sciences of the United States; he also received the Wolf Prize of Israel and the Albert Lasker Special Public Health Award, among other distinctions.

In the book *Nobel: The Man and His Prizes,* the Nobel Foundation stated that "it is to be regretted" that the discovery of DNA being the transforming principle was not awarded the Nobel Prize.[3] It was a rare admission of a mistake from the Nobel Foundation!

We recorded our conversation in 1997 in McCarty's office at Rockefeller University.

About Oswald T. Avery.

"He was not a very outgoing person. He was a small man who was quite restrained, at least by the time I got to know him. He was a lifelong bachelor. At the time I knew him, he no longer liked to talk in public. We induced him to talk at our regular staff meeting in December 1943. By then, our paper was in press, but he had not talked there for years. He was President of the Society of American Bacteriologists the year that I came to Rockefeller. He gave the Presidential Address, and he would not let it be published. Talks of this kind were not science, and he just didn't want his general comments in print.

So he was known as a reserved person, somebody who would not rush to publish.

There is no doubt about that....

Was Avery's being so reserved in disseminating the discovery frustrating for you?

Obviously, there was some frustration, but he was a very likable person and quite revered around here; and everybody looked out for him, particularly since he had been ill. He had hyperthyroidism, a disease in which the thyroid overworks. They had to operate and take out a considerable part of his thyroid gland. He was ill for years, and people were very solicitous about it, including MacLeod. Avery had a tremor, which comes with this illness, he could not do experimental work anymore, and he got quite depressed at times. He was just recovering fully in the late 1930s."[4]

---

[3] The Nobel Foundation and W. Odelberg, eds. *Nobel: The Man and His Prizes,* 3rd ed. (Elsevier: New York, 1972), 201.
[4] *Candid Science II,* 29.

MATTHEW MESELSON

We should be absolutely opposed to biological weapons.

Matthew Meselson in 2004 in Woods Hole. (photograph by I. Hargittai)

Matthew Meselson[1] (1930–) was born in Denver, Colorado. He is the Thomas Dudley Cabot Professor of the Natural Sciences in the Department of Molecular and Cellular Biology of Harvard University in Cambridge, Massachusetts. He did his graduate work with Linus Pauling at the California Institute of Technology. While he was a graduate student, he and Franklin Stahl devised an experiment in which they provided proof for the semiconservative replication of DNA in 1957 and published it in 1958. Their experiment employed a new technique—which they invented—of using centrifugal force to separate molecules based on their densities. It is called the *density gradient centrifugation*.[2] We recorded our conversation in 2004 in Woods Hole, Massachusetts.

---

[1] I. Hargittai and M. Hargittai, *Candid Science VI*, 40–61.
[2] Their joint work was described by Frederick L. Holmes in *Meselson, Stahl, and the Replication of DNA: A History of "The Most Beautiful Experiment in Biology"* (New Haven, CT: Yale University Press, 2001).

We knew that Meselson had been interested in bioterrorism since 1963 and had also heard that Linus Pauling had told him that he should first establish himself as a scientist, and only then turn to such questions. We asked him about that.

"That's right. What happened was—and I only learned about this later—there was this new organization, the General Advisory Committee of the United States Arms Control and Disarmament Agency. They had too much money in their first year, and they didn't want to give it back because in the next year maybe they would not be given as much money by the office of the budget. They were a brand new agency and they were expanding, so it was wise of them not to have a surplus. They got the idea to invite six academics for the summer and pay them. My officemate was Freeman Dyson, which was wonderful for me because he encouraged me to trust my intuition.

They said that we should work on European nuclear-theater nuclear weapons arms control. For about a week or two I tried that, but it became quickly obvious that I didn't know anything about it, which is OK; but a lot of other people, like Henry Kissinger, were experts in this. Kissinger had written a book about the necessity of choice. So I asked my boss if, instead, I could work on biological and chemical arms control. My boss was Frank Long, a wonderful chemist, a professor at Cornell University in Ithaca. He said, sure, I could work on them. He also told me that there was a man before we came who worked on biological weapons arms control, but he became so depressed that he killed himself. Long offered me his desk. He also said that they were all going for Moscow to negotiate the nuclear test ban treaty; so he gave me complete freedom to do what I wanted.

I decided to start with biological weapons because chemical and biological weapons were too big a subject. I went to the CIA to see what other countries were doing at that time. The answer was that we didn't know much. It's probably still true. I also went to find out what we were doing. We had a big, offensive, biological weapons program. A very nice man named Leroy Fothergill was my guide. I had to get all kinds of security clearances. I asked him why we were doing this, working on biological weapons. He told me that it was much cheaper than all kinds of nuclear weapons. I was thinking about what he told me. It puzzled me because we had all those nuclear weapons. Why would we want to pioneer weapons of mass destruction so cheap that everybody could have them? I found this crazy. We should be absolutely opposed to biological weapons. We should be the last country to make such weapons. If we set an example, everybody else is going to make them. Rather, we should try to make them illegal; we should get rid of them.

At first, I didn't say much in public. I didn't write because it was better if nobody talked about such things. You don't want to get people interested in them. But then the Vietnam War came and people were accusing us of waging chemical warfare in Vietnam. There was more talk about it. There were some accidents of various kinds that came to public notice. I became more public about it. Then, when President Nixon was elected,...he chose Henry Kissinger to be his National Security Advisor. I knew Henry because his office building was next to the Biological Laboratories at Harvard University. I went to his arms control seminar and I would also see him occasionally at lunch. For some reason, we did become friendly. We were both going through a divorce at the same time. For us both, Harvard was an emotionally frigid environment. I remember when Henry...came back from his second trip to Vietnam—he was not yet in government; it was under President Johnson, and he had gone there at the request of the American Ambassador, Cabot Lodge—he was very tired. We were sitting on the sofa in his office and had some sherry. He said, 'Now I know how the good Germans felt.' This really touched my heart. I thought that the Vietnam War was a mistake, and he wouldn't say that.

Did he mean by 'good Germans' those who were against Hitler?

He meant by good Germans those who saw what was going on. It didn't occur to me until you just said it that that could have been his implication. But I think he meant only the Germans who saw that their country was doing something wrong.

They saw and did nothing.

I don't know how to answer that."[3]

---

[3] *Candid Science VI*, 49–51.

CÉSAR MILSTEIN

I think it [cloning humans] is horrible. It would be a terrible stupidity.

César Milstein in 2000 in Cambridge, UK. (photo by I. Hargittai)

César Milstein[1] (1927–2002) was an Argentinean-born (in Bahía Blanca) British molecular biologist. He started his college education at the Colegio Nacional in Bahía Blanca and completed it at the Universidad de Buenos Aires with a degree Doctor en Química in 1957. He started his career in Buenos Aires and continued in Cambridge, where he earned his PhD degree and became one of the leading scientists of the MRC Laboratory of Molecular Biology (LMB). In 1984, Milstein was awarded the Nobel Prize in Physiology or Medicine "for theories concerning the specificity in development and control of the immune system and the discovery of the principle for production of monoclonal antibodies"[2] together with Niels K. Jerne and Georges J. F. Köhler. We recorded our conversation in 1998 and augmented it in 2000, at the MRC LMB.

---

[1] I. Hargittai and M. Hargittai, *Candid Science II*, 220–37.
[2] "The Nobel Prize in Physiology or Medicine 1984," *Nobelprize.org*. Nobel Media AB 2013, http://www.nobelprize.org/nobel_prizes/medicine/laureates/1984/.

César Milstein with his father in 1984 in Stockholm. (courtesy of César Milstein)

The photograph of Milstein with his father suggests a special relationship between them, and he told us about it.

"My father came [to Argentina] from the Ukraine when he was fourteen. He had no family support and became a farm laborer and somehow through his own efforts learned to write and read Spanish. Over the years he tried a variety of jobs (carpentry, rail-worker, etc.) and became active in cultural activities involving Yiddish libraries and nonreligious Jewish organizations, some with anarchist or anarcho-sindicalist connections. Eventually, he started as a shop assistant and later became a traveling salesman. My mother was born in Argentina of a family recently arrived from Lithuania. She was a very gifted child, and with great sacrifices, went to secondary school and eventually became the head teacher of a primary school when still very young. We were three brothers, and for both of my parents their highest aspiration was that we all went to university. All of us graduated.

My parents were very strict, and I was very rebellious, and on many occasion our relations became extremely tense. Somehow, the conflicts were kept at a level that never eroded the basic love and appreciation that I had for them. When my parents realized that I was more interested in an academic career than in the professions, they were incredibly supportive. My mother typed my Argentinean PhD thesis; and my father repeatedly offered me economic assistance so that I could dedicate full time to my research. I refused (except for the down payment of a mortgage that was their wedding present). I was determined to feel fully independent even though in those days there were no scholarships in Argentina for a research student and I had to do part-time work (as did my wife) so that we could support ourselves. I dedicated my Argentinean PhD dissertation to them, and they (especially my mother) became extremely emotional at the gesture. As time went by, they understood and respected my demands for total independence and our relationship quickly lost all the rough edges. They were sad when I told them that we were immigrating to Britain, and it was first my mother who came to accept and support our decision.

They were both thrilled and proud to learn of my steady progress, and I made efforts to visit them in Argentina at least once a year. Unfortunately, my mother died a few years before my first big jump in academic recognition (the Fellowship of the Royal Society [FRS]). My father, however, came to London to the ceremony. He was—in his tranquil and reflective manner that clearly contrasted with my mother's emotional personality—obviously proud but also somewhat bewildered and amazed that this was happening to him. He also came with me to Jerusalem when I received the Wolf Prize. On that occasion, however, he could not stop himself when I gave my three-minute speech of thanks. I mentioned that I was a typical example of a Jew of the Diaspora, the beneficiary of the determination of parents that were willing to make all sorts of sacrifices to see their children receive higher education. At this point, he jumped from his seat, came to the rostrum (the ceremony took place at the Knesset and I received the award from the hands of the President of Israel), and to my great embarrassment, gave me a kiss! The audience, however, loved it!

By the time he came to Stockholm, he was more prepared for such occasions and while very moved, he was also very composed when he gave with me a live interview to an Argentinean TV station. By then, he had already been interviewed on several occasions in Argentinean TV. He was a natural storyteller and the interviewers loved him for that. Subsequently, and on several occasions over the following two years or so, he was interviewed for his knowledge of the Jewish immigration to Argentina in the beginning of the past century. He was particularly good at telling the story of a poor Jewish immigrant's struggle in a foreign land, determined not only to raise a good Jewish family but also to contribute to the maintenance of the Jewish culture and the Yiddish language. It gave me great happiness to feel that somehow I was repaying all his own efforts by bringing joy and a new interest to the last years of his long life."[3]

---

[3] *Candid Science II*, 235–37.

SALVADOR MONCADA

The best way to run an institute is not to try to run it.

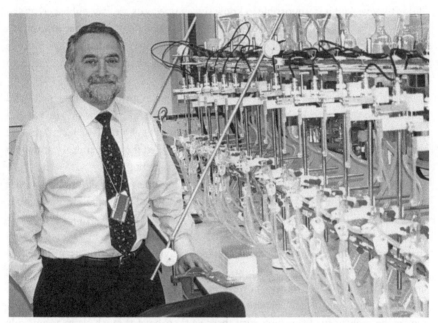

Salvador Moncada in 2000 at University College London. (photo by I. Hargittai)

Salvador Moncada[1] (1944–) is a Honduran-born (in Tegucigalpa) British biomedical scientist. He graduated as doctor of medicine and surgery from the School of Medicine, University of El Salvador, in 1970. He received his PhD degree in pharmacology from the Royal College of Surgeons, University of London, in 1973. He worked for a few years each in El Salvador and in Honduras; and between 1975 and 1995, for the Wellcome Research Laboratories (UK), rising to be their Director of Research. He has been at University College London since 1995. He is the Director of the Wolfson Institute for Biomedical Research.

Moncada's two main research areas have been prostaglandins and nitric oxide. He is most famous for having identified the so-called endothelium-derived

[1] I. Hargittai and M. Hargittai, *Candid Science II*, 564–77.

relaxing factor (EDRF) as nitric oxide (NO). When the 1998 Nobel Prize in Physiology or Medicine was announced and Moncada was not among the three awardees, an international uproar followed in which it was suggested that the Nobel Committee was rewriting science history. Otherwise, Moncada has been recognized by high positions and prestigious awards and academy memberships.

We recorded our conversation in 2000 in Moncada's office at the Wolfson Institute, University College London.

We asked Moncada to reflect on what may have happened with the 1998 Nobel Prize.

"I don't know. And it is impossible to know since no records are ever published on the reasons why a decision is made. So it is useless to speculate. I think that institutions can give prizes to whomever they want, that is their prerogative. The problem with the Nobel is that it has become so preeminent that it tends, when mistakes are made, to 'rewrite history.'

I have said before that I am extremely proud of our contribution to the birth and development of this field. Our work on the identification of NO was the earliest and has been recognized as such by the scientific community at large. The techniques we developed for this are now universally used. We identified the biochemical pathway to the synthesis of NO and made many more seminal contributions.

Now, when a scientific discovery is made it illuminates not only lines for future research but also the past and things, which were scattered in the literature, suddenly come together. That, however, is hindsight!

Let me give you an example. In the 1970s, a Japanese group led by Deguchi, found that L-arginine stimulated the enzyme, soluble guanylate cyclase. This is probably the closest work to the discovery of the L-arginine→NO pathway, which is the discovery we are talking about. Those workers, however, were not aware of the great significance of their work. This only became apparent some ten years later. However, if we are to recognize the prehistory of this discovery, these workers have as much claim as anybody else. The history of science is full of incidents which precede a discovery and can be connected with it only after the discovery takes place.

Having said that, I know of no case in which such a significant group of the scientific community has protested so vehemently about the unjust exclusion of a scientist. This is reassuring. I am very happy about the recognition, which I have received through peer review, the most objective form of recognition. On the 1998 decision about the physiology or medicine prize, the Nobel committee was very much at odds with the scientific community."[2]

About his background and education.

"I come from a mixed background. I was born in Honduras but my mother came from Central Europe. She was born in a small town in northwest Romania. Her name was

---

[2] Ibid., 566.

Jenny Seidner and she was of German Jewish stock. My maternal grandfather was an Austrian Jew who fought in the First World War. My mother died some years ago. My father, on the other hand, is a Central American of Spanish descent, most probably Catalan, since my surname, Moncada, comes from that region. So I grew up in a mixed Catholic and Jewish household, but since my father was not a strong Catholic, the Jewish religion predominated. I went many more times to the synagogue than to the church, until the age of thirteen, when I decided not to do the Bar Mitzvah. At that point, I abandoned religion for many years, until, about twelve years ago when, being by chance in Amsterdam on the date of the Jewish New Year, I decided to visit the old synagogue out of curiosity. The service was very moving. It touched me deeply, more so when I realized that a number of people attending had Spanish or Portuguese surnames. They were obviously from old Sephardic families coming from the Iberian Peninsula since the sixteenth century! I did not become religious suddenly; however, it highlighted for me my cultural heritage. I enjoy very much, when I go to the United States, to visit my sister in New York and dine with her and her family for Sabbath. I feel at home. I don't pray but, somehow, I feel extremely comfortable in that environment.

For a variety of circumstances, which would take too long to tell, my grandfather and his family, who left Europe in 1937, did not go to the USA or Argentina or any of the other places that Jews were emigrating to. Instead, they ended up in Honduras! I was born there, but my family moved to El Salvador when I was about four years old. I grew up in El Salvador and did all my studies to obtain a medical degree in 1970. Fortunately, the Medical School in El Salvador was, at that time, very good. A group of highly talented academics, with financial help from the Kellogg and Rockefeller Foundations, had modernized medical education in El Salvador, so I was exposed to high quality training. However, because of my heavy involvement in student politics during my medical studies, I was persecuted and finally expelled from El Salvador. This was shortly after I finished my medical education, and I ended up back in Honduras. From there, I decided to look for a place for postgraduate training."[3]

---

[3] Ibid.,567.

BENNO MÜLLER-HILL

We should distinguish between knowledge and what you do with this knowledge.

Benno Müller-Hill in 1999 in Cologne, Germany. (photo by I. Hargittai)

Benno Müller-Hill[1] (1933–) was born in Freiburg im Breisgau, Germany. He is a biochemist and geneticist and Professor Emeritus at Cologne University, and he made important contributions to genetics by isolating the first transcription factor, the *lac* repressor. He wrote an unusual book about it.[2] He is also well-known for his research into the history of genetics in Nazi Germany, which is infamous for heinous crimes against humanity, and for the silence surrounding its history in postwar Germany as well as in today's Germany. Müller-Hill's book about it appeared in German under the title *Tödliche Wissenschaft*, and later, in English as *Murderous Science*.[3] We recorded our conversation in 1999 at the University of Cologne.

---

[1] I. Hargittai and M. Hargittai, *Candid Science II*, 114–29.
[2] B. Müller-Hill, *The lac Operon: A Short History of a Genetic Paradigm* (Berlin and New York: Walter de Gruyter, 1996).
[3] B. Müller-Hill, *Tödliche Wissenschaft* (Reinbek: Rowohlt Taschenbuch Verlag, 1984). Translated by George R. Fraser as *Murderous Science: Elimination by Scientific Selection of Jews, Gypsies, and Others in Germany, 1933–1945* (New York: Cold Spring Harbor Laboratory Press, 1998).

We asked about the reaction to his book Tödliche Wissenschaft.

"In Germany, silence."[4]

We asked him if he would mind being remembered more for his anti-Nazi studies than for his genetics.

"I would hope that the isolation of the *lac* repressor will also count. When I came to Wally [Walter] Gilbert's lab, it seemed to be a solvable problem, and I wanted to solve it very much. It also seemed to be a problem where you could put in all your efforts with a fair probability of failure. Normally, when you do science, even if you don't achieve your original goal, something else may come out of your efforts, which you can publish. In this case, however, it seemed to be all or nothing. This was so not only for me but for Wally as well. At the time, he was teaching theoretical physics and he didn't have tenure at Harvard. So we were both having high stakes in this project. He didn't have many publications because he followed Jim [James D.] Watson's policy of not putting his name on a paper unless he was involved in the experimental work. He had about eight graduate students and he was showing them how to do the experiments but he was never a coauthor. This is very good for the graduate student, for developing independence and responsibility, but may not be very good for the professor's formal productivity.

Doesn't this approach downplay the intellectual contribution?

It does. The intellectual contribution disappears, and what counts is only what you do. On the other hand, this approach also implies that if you have a really good idea, you should also do it.

Wally and I talked a lot but didn't really know how to attack the problem. Then one day somebody came from Paris and told us that he'd solved the problem. Actually, he did not but for us it had a shocking effect.

The *lac* repressor was a notion and we wanted to isolate it and have the actual molecule in our hands. Supposedly, it was a protein in extremely low concentrations. My idea was that we could use a particular mutant of the gene, which produces this repressor. Finally, I produced this mutant, Wally tested it, and it worked but it was like walking on thin ice. Other people tried to get it too but, fortunately, they did not succeed before we did. Our first publication appeared in December 1966 in the *Proceedings of the National Academy of Sciences*. I've never lived in such an atmosphere of excitement and hard work ever since. We gave our utmost to the project.

On the one hand, there was Wally with whom I could discuss the experimental details, and he loved to talk things over. On the other hand, there was Jim Watson, who didn't talk at all. When he came into the lab, you would say one or two sentences, and that was it. You just realized that there was nothing more to say. For example, once I got a particular mutant and I was very proud of that, but you can explain such an experiment in two minutes. There was another German postdoc there whom I knew from Freiburg, Klaus Weber. We had a bet about who spoke how much with Jim. After half a

---

[4] *Candid Science II*, 120.

year, he was at twenty-two minutes and I was at seventeen minutes, total. Jim Watson's nonspeaking was even more driving. It had the effect that there was someone with whom you could speak only if you had something to say. If you had no result, there was nothing to speak about.

Listening to your description of your work on identifying the lac repressor at Harvard, I feel an even greater appreciation for your tremendous work and self-sacrifice in uncovering the history of Nazi science.

I would like to simply say that finding these records and writing on them was equivalent to what I did with Wally Gilbert. I found something unique and I had to do it.[5]

As a geneticist, how do you view the relationship between genetics and intelligence. It seems to be a taboo topic.

You're not supposed to talk about it. There's a total misunderstanding about it even among scientists, which is the following. If you could show that there are genetic differences related to intelligence and that some ethnic groups have more favorable genetics for intelligence than other groups, or the other way around, this would, per se, invite the accusation that you are attacking the losers. This is a total mistake. We have to face reality that such differences may exist but you should never use the information about such differences for the purpose of attacking people. We should distinguish between knowledge and what you do with this knowledge.

The main problem is that you have to have a hard definition of what is called intelligence. Great difficulties begin here. The psychologists and the psychiatrists are simply unable to deliver really hard definitions. Since there is no such definition, you cannot do genetics."[6]

---

[5] Ibid., 127–29.
[6] Ibid., 127.

PAUL NURSE

My parents could never understand what I could possibly be doing by still learning in my mid-thirties.

Paul Nurse in 2003 in Budapest. (photo by I. Hargittai)

Paul M. Nurse[1] (1949–) was born in London. He received his undergraduate degree (1970) and his PhD degree (1973) from the University of Birmingham and the University of East Anglia, respectively. In 2001, Nurse shared the Nobel Prize in Physiology or Medicine with Leland H. Hartwell and R. Timothy Hunt "for their discoveries of key regulators of the cell cycle."[2] Nurse has received other top awards and filled important science-administration positions, including the presidency of Rockefeller University (2003–2010). Currently, he is President of the Royal Society (London) and the chief executive of the new Francis Crick Institute, which, when it opens in 2015, will be the largest biomedical research center in the UK. We recorded our conversation in 2003 in Budapest.

---

[1] I. Hargittai and M. Hargittai, *Candid Science VI*, v–vi, 62–87.
[2] "The Nobel Prize in Physiology or Medicine 2001," *Nobelprize.org*. Nobel Media AB 2013, http://www.nobelprize.org/nobel_prizes/medicine/laureates/2001/.

On the importance of understanding how science and scientists work.

"Why is it important to understand how science and scientists work? The characteristics of science, respect for observation and experimentation, reliance upon refutable hypotheses, consistency across different fields of inquiry, critical skepticism, all contribute to making science the most reliable approach for gaining knowledge about the natural world. Advances in science have spectacularly changed our understanding of the world and of what it is to be human, and have also underpinned many new practices and technologies which have improved the quality of human life. This has happened to such an extent that it is now impossible to imagine society without the benefits that science has provided. Although science has this universal relevance, scientists are trained specialists and are relatively few in number, and there is always a danger that they can easily become remote from the rest of society."[3]

On manners and style in science: scientific discovery versus artistic creations.

"I always find this a little difficult to deal with. There is some truth in the fact that the discovery of something is inevitable to happen; if A won't do it then B will do it. But I would like to make a few observations about it. First of all, the manner in which a discovery is done, the experiments and the thinking, can be very different. The manner, in which the discovery is unraveled, reflects creativity, different approaches, personalities in the same way as the work of great literature may be revealing some truth about human nature. That truth, that universal message, may be also in another work of literature of a completely different form, but what resonates with you is the message. Of course, while a piece of literature is completely unique and different, but still the universal messages may have great similarities. Science is not quite as unique, but although the underlying story to be discovered may be similar and different people may be out to approach it, the way in which they do it and the beauty with which it is done, with the risk of sounding pompous, differs. Then there is the question of timing, and we shouldn't forget about it, there are discoveries, which can be made now or can be made in thirty or fifty years. Often things come together at once because of the set of conditions that lead to certain problems being addressed and having the methodologies to do them. But that is not always the case. Some of the problems that we think about now, I remember that we discussed them as graduate students thirty years ago. We thought a lot about them, did some work, could not solve them and now we're coming back to them. It is true that there are some basic structures out there to be discovered, but it's more complicated than that. The way with which we approach them, the timing by which they are sorted out depend on the individuality of the scientists involved."[4]

---

[3] *Candid Science VI*, v–vi.
[4] Ibid., 84.

On his career.

"My career has been quite purposeful; I used genetics to find out what elements were important for control; then I translated that into a molecular role, and then I showed the validity of my findings for other organisms. It was a rather clear pathway through. I wasn't propelled to the front. Other people are propelled to the front. They're working always at the leading edge, in the limelight. It's partly because they keep an eye on the moving front and go to it. There are then others who are in the forefront some times; they are working somewhere else and find themselves suddenly propelled to the forefront. There are different ways of doing things and different scientific tastes."[5]

---

[5] Ibid., 85.

CHRISTIANE NÜSSLEIN-VOLHARD

When a woman is aggressive and behaves in the same situation as a man would, they always charge that against her.

Christiane Nüsslein-Volhard in 2001 in Tübingen, Germany. (photo by M. Hargittai)

Christiane Nüsslein-Volhard[1] (1942–) was born in Magdeburg, Germany, and received her education and doctorate in Tübingen. She is a developmental biologist at the University of Tübingen. In 1995, Nüsslein-Volhard was awarded the Nobel Prize in Physiology or Medicine together with Edward B. Lewis and Eric F. Wieschaus "for their discoveries concerning the genetic control of early embryonic development."[2] We recorded our conversation in 2001 in Nüsslein-Volhard's office at the University of Tübingen.

---

[1] I. Hargittai and M. Hargittai, *Candid Science VI*, 134–51.
[2] "The Nobel Prize in Physiology or Medicine 1995," *Nobelprize.org*. Nobel Media AB 2013, http://www.nobelprize.org/nobel_prizes/medicine/laureates/1995/.

Nüsslein-Volhard on whether she experienced gender discrimination at various stages of her career.

"Oh, yes, plenty! But probably it is better not to dwell on it too much. First, we can go back to the time when I grew up. The general problem then was that women simply were not considered to be professionals; professionals enough to have big important jobs. Therefore, they were just often overlooked and no one entrusted them with important things. I have to say that my science was never discriminated against. So I had no problem whatsoever in getting my science recognized. But in the practical aspects, to get jobs, to get money, to get lab space, women have not been treated equally, I think.

At the time when I was a young scientist, men often had a family and children, and they got the better positions automatically. The professors always said, 'But this is a man, he has to support and nourish a family, so this is why he is going to get the job and you will not.' This happened repeatedly to me. The same thing with promotions. They often said, 'You are a woman, and there is a man and he deserves it more.' Actually, my worst experience happened with my PhD; I collaborated with a man, I did most of the work, I wrote the paper and then he got first authorship because my boss said, 'He has a family and for him this is important for his career. You are a woman and married, so it does not matter.' This was particularly unjust because he gave up science right after his thesis and went to teach where he did not need the publication at all. Whereas, I did suffer in trying to find jobs later because I did not have this publication with my name as a first author. So discrimination started right away, at the very beginning.

When I did my diploma thesis in this institute, it was customary that the diploma students got some stipends, but the big boss said, 'She does not need it, she is married.' This kind of thing is over by now. But the old professors still think that it is totally legitimate to pay men more if they have families to support. This still happens. If there is a single woman who is good and does a perfect job and there is a man who is mediocre but has a wife and family, he will get the promotion.

Nowadays special attention is paid to this question, and often there is a 'positive discrimination' in that even if a woman is not as qualified as a male candidate, she may get the job.

Yes. Some women say that it happened so long and so often the other way around, why not to have some advantage of this, for a change. I do not agree with that. However, it turned out that when you push the women a little bit more and you get more women in positions, they all turn out to be better than you have thought.

In the Max Planck Society, they created some jobs specifically for women, and they also raised the percentage of women at the independent group-leader level dramatically in all their institutes and filled these jobs in a short period of time. It turned out that all these women are very good in their jobs. This says that these women were simply overlooked previously. It seems that you can do a little push without the danger that...women will not do well. Generally, they had been so underrated that it was good to give them better status.

Do you think that a woman has to work harder to get the same recognition as a man?

At least she can't afford to make mistakes, at least not as much as men. When a young man does that, they say, 'oh, he is ambitious' and 'it can happen to anyone.' But when a young woman does that, it is usually overrated. In particular, when women are aggressive, people don't like that at all. When men are aggressive, it is perfectly normal;... when a woman is aggressive and behaves in the same situation as a man would, they always charge that against her. This is bad because aggression in a positive sense is absolutely necessary to succeed in a job. You have to push your point, and not that of your neighbor. Unfortunately, it is part of the profession.

Do you think that if a woman has a family, it is a disadvantage for her in science?

It could also be an advantage, I don't really know. I think it does not really matter very much; at any rate, it matters less than people think. When you have a husband and you have two salaries, you can afford to hire a nurse or household help; you can afford day care, and then life is much easier than when you are living alone and doing all the work in your household. I don't know.

Do you have any advice for young women who want to have science careers and also a family?

Just go for it, and try to find professional help for your household and for day care as soon as possible and don't be stingy on that. Get the best help, pay a lot of money to the best servants and let them do the job for you. Many people do not do that; they just try to do both jobs and, of course, then they crash.

The few high-positioned women are often looked for in committees, public appearances, and so on. Does this make it hard to keep your eye on the ball?

Oh, yes. It is very hard. There is also this women's issue on top of it; it is hard to deal with all of them. I am on many committees. I am on committees for prizes, for founding institutions, for hiring people, for European commissions on ethics, and so on. This is a lot or work. I also sometimes appear on TV about science."[3]

---

[3] *Candid Science VI*, 145–47.

MAX F. PERUTZ

I really owe my scientific career to his [W. Lawrence Bragg's] support.

Max F. Perutz in 2000 in Cambridge, UK. (photo by I. Hargittai)

Max F. Perutz[1] (1914–2002) was an Austrian-born British crystallographer and molecular biologist. He embarked on a detailed structure elucidation of globular proteins when the experts found it questionable whether such task could be solved. Perutz applied techniques of the isomorphous replacement method, which had been used in small-molecule crystallography, and successfully attacked protein structures. Perutz and his former student John Kendrew received the 1962 Nobel Prize in Chemistry "for their studies of the structures of globular proteins."[2] We recorded our conversation in 1997 in Perutz's office at the MRC Laboratory of Molecular Biology in Cambridge.

---

[1] I. Hargittai and M. Hargittai, *Candid Science II*, 280–95.
[2] "The Nobel Prize in Chemistry 1962," *Nobelprize.org*. Nobel Media AB 2013, http://www.nobelprize.org/nobel_prizes/chemistry/laureates/1962/.

James D. Watson, Max F. Perutz, César Milstein, Frederick Sanger, John Kendrew, and Aaron Klug, all Nobel laureates of the MRC Laboratory of Molecular Biology. (courtesy of the MRC LMB Archives, Cambridge, UK)

Initially, Perutz's group published a large number of erroneous models for protein structures.[3] In what seemed to be a race, Linus Pauling defeated Perutz when he announced his alpha-helix model. Pauling's superiority stemmed from his thorough knowledge of structural chemistry that Perutz apparently lacked. Perutz must have thought a great deal about this gaffe; he gave this impression when he responded to our question about this story.

"The awful thing is that Kendrew and I didn't know that the peptide bond was planar. Perhaps the most important observation that misled us was the X-ray pictures of alpha-keratin by [William] Astbury at Leeds. Astbury discovered that protein fibers gave two kinds of pictures. Wool gave a picture with 5.1-Å meridional reflection, but when he stretched wool in steam, that disappeared and a reflection at 3.4 Å appeared instead. Astbury argued rightly that the stretched wool corresponds to an extended polypeptide chain and the 5.1 Å to a folded one. So Kendrew and I thought we must build molecular models with a repeat of 5.1 Å, and we built a variety of helical structures and nonhelical structures, all with that repeat. We noticed that we had to strain our models rather to get that repeat, but we didn't see any way out. And we didn't realize that we had to keep the peptide bond planar. If we had, our model would have been even more strained. We would have had to force it even more to achieve the 5.1-Å repeat. Pauling, on the other hand, disregarded Astbury's pictures; he built a model

---

[3] W. L. Bragg, J. C. Kendrew, and M. F. Perutz, "Polypeptide Chain Configuration in Crystalline Proteins," *Proc. Roy. Soc.* 203A (1950): 321–357.

that was stereochemically reasonable. He kept the peptide bonds planar and made the best possible hydrogen bonds.

One Saturday morning, I went to the Cavendish Library and found this series of papers by Pauling and saw his structure, and it looked stereochemically right, and yet it didn't seem right because it didn't have the right repeat. My mind was in turmoil and I kicked myself for having missed this. I cycled home to lunch and my family wondered why I didn't listen to anything they said. Suddenly a thought occurred to me. If there was really the regular axial repeat of 1.5-Å of the residues along the Pauling model which he indicated, there must be a reflection at 1.5 Å on the meridian which has not been reported. I remembered that I had a horse hair in the drawer in the lab so I cycled back to the lab.

There was also another thing. I had visited Astbury's laboratory and seen his setup. He had a goniometer head and behind it the photographic film, which was quite narrow. If there were a 1.5-Å reflection, the Bragg angle would be 31° and the reflection would occur at an angle of 62° from the incident beam. I realized that Astbury's plate would have missed that. Moreover, Astbury always had the fiber axis at right angles to the X-ray beam and I realized that to get that reflection, one would have to incline the fiber at 31° to the incident beam to fulfill Bragg's law.

So I went back to the lab, found my horse hair, set it up at an angle of 31° to the beam, and put a cylindrical film around it, so that it would catch reflections at high angles; I exposed it for two hours, and developed the film with my heart in my mouth. And to my surprise, what I found was a strong reflection at 1.5 Å which, I realized, excluded all models except the Pauling-Corey alpha-helix.

What next? That was on a Saturday. On Monday morning I rushed into Bragg's office and told him what I'd discovered. Bragg asked, how did you think of that? I told him it was because I was so angry that I hadn't thought of that structure myself. To which Bragg replied coldly, I wish I'd made you angry earlier.

You did your experiment on Saturday and you rushed into Bragg's office on Monday. How could you wait so long? Why didn't you let him know during the weekend?

Relations were a little more formal then than now. I wouldn't have disturbed him on Sunday at home."[4]

---

[4] *Candid Science II*, 287–88.

FREDERICK C. ROBBINS

From the day of the [Nobel Prize] announcement, I never got a request to see a patient in consultation.

Frederick C. Robbins in 2000 at Case Western Reserve University in Cleveland, Ohio.
(photo by M. Hargittai)

Frederick C. Robbins[1] (1916–2003), an American pediatrician, was born in Auburn, Alabama. He received his doctor of medicine degree from Harvard University in 1940. From 1942 to 1946, he served in the US Army. He spent his residency at the Boston Children's Hospital, where he joined John F. Enders and became involved in polio work. Enders, Robbins, and Thomas Weller shared the 1954 Nobel Prize in Physiology or Medicine "for their discovery of the ability of poliomyelitis viruses to grow in cultures of various types of tissue."[2] Their work made it possible for Jonas Salk and Albert Sabine to (independently) develop vaccines against poliomyelitis, which eventually lead to

---

[1] I. Hargittai and M. Hargittai, *Candid Science II*, 498–517.
[2] "The Nobel Prize in Physiology or Medicine 1954," *Nobelprize.org*. Nobel Media AB 2013, http://www.nobelprize.org/nobel_prizes/medicine/laureates/1954/.

Albert Sabin and Jonas Salk on US postage stamps (2006).

the almost total eradication of polio all over the world. We recorded our conversation in 2000 in Robbins's office at the Department of Epidemiology and Biostatistics of Case Western University in Cleveland, Ohio.

This is how Robbins got into polio research.

"My father was a scientist, a botanist. I often worked in his laboratory as a boy, washing dishes. That did not really turn me on. Also, just like many young people I did not want to be just a copy of my father, I wanted to be myself.

I had first attended the University of Missouri Medical School, but there was only a two-year training program there. Then I transferred to the Harvard Medical School. There I saw faculty members who were clinicians, but they were also scientists—and that intrigued me. When I received my medical degree, I went to the Children's Hospital for my post MD training; those of us who came directly from medical school had to spend a year in the laboratory. I was assigned to the bacteriology laboratory. Thus, when I went into the Army in 1942, I was assigned to laboratory service and given the responsibility for the Virus and Rickettsia Laboratory of the 15th General Medical Laboratory of the US Army. Of course, I did not know much about viruses and very little about rickettsia. I was sent to Italy, where I stayed for two years. I did studies on Q-fever and hepatitis, and that increased my interest in research and infectious diseases. When I came back from the war, I decided to do research and teaching.

I started to be interested in virology already a little earlier because my roommate in the fourth year of medical school, Tom Weller, chose to do elective work with Dr. John Enders, who was studying viruses. Tom kept telling me about the interesting work they were doing and what a wonderful person Dr. Enders was, so that initiated my interest in viruses.

After the war I completed my residency, and then I joined Dr. Enders. I received a three-year fellowship, and the original plan was that I spend the first year...with Dr. Enders and two years with Dr. Macfarlane Burnet in Australia. Since at the end of the first year we started to get very interesting results on polio, Dr. Enders suggested that I not go to Australia but stay with him.

I started out to see if I could use tissue culture to cultivate a virus from infant diarrhea—in those days infant diarrhea was a big problem. Dr. Weller, who preceded me in Dr. Enders' laboratory, was trying to cultivate the virus of chickenpox. As I remember—Dr Weller does not remember it quite this way—we were talking about the fact that polio was predominantly an infection of the intestines. Dr. Enders suggested that, since we had some tissue cultures, why don't we just put some polio in? So this is how it started."[3]

The Nobel Prize changed Robbins's life.

"Everybody started to expect things of me that I could not deliver. One of the local science writers from a local paper here, for example, came to me about six months after the prize and asked, 'What have you done now?' There was another strange outcome that I cannot quite explain. From the day of the announcement, I never got a request to see a patient in consultation. The Prize also meant that I was more in demand for various committees and that sort; so it was somewhat distracting. To some extent it increased the pressure and my expectations of myself. For a certain length of time I was very disappointed in my own performance, but finally I adjusted to it."[4]

---

[3] *Candid Science II*, 500–502.
[4] Ibid., 513.

JENS CHR. SKOU

I would not work under conditions where you need to apply for grants to buy a pencil.

Jens Chr. Skou in 2003 next to the sign for Jens Chr. Skou Street in Aarhus, Denmark. (photo by I. Hargittai)

Jens Chr. Skou[1] (1918–) was born in Lemvig, Denmark. He received his doctor of medicine degree in 1944 from the University of Copenhagen. He participated in clinical training but eventually opted for a career in research, and he remained affiliated with the University of Aarhus, Denmark, all his life. In 1997, Skou was awarded half of the Nobel Prize in Chemistry "for the first

---

[1] B. Hargittai and I. Hargittai, *Candid Science V*, 428–53.

discovery of an ion-transporting enzyme, Na$^+$,K$^+$ATPase."[2] We recorded our conversation in 2003 in his office at the University of Aarhus.

Skou was finishing medical school at the time of the German occupation of Denmark.

"During the war we lived in the atmosphere of the occupation and it had an impact on all of us. In my medical student group we knew that one of the students was an informer for the Germans. He was liquidated. Because of fear of revenge from the Gestapo against the group, teaching was cancelled. Many of our teachers who took part in the resistance against the Germans disappeared, went underground or escaped to Sweden. We managed to get our final exam in the summer of 1944... But we did not dare to assemble to sign the Hippocratic oath, but came one by one to a secret place. I started my internship at a hospital in the northern part of the country. In the surgical ward, the head of the department had escaped from the Gestapo to Sweden. The next in line in the department was anxious to teach me how to operate for appendicitis and alike. This was unusual for someone who had just started on his internship, but pleased me because I intended to become a surgeon. Eventually I realized that the reason for this was that he was involved in receiving weapons from England delivered by plane at night. When we were on night duty together and he had to leave to receive weapons he wanted to be sure that I could take over. He was later arrested by the Gestapo, but he survived."[3]

About the funding system of scientific research.

"In general I am very skeptical about funding systems where you have to find all the money for your research from funds. To write an application for a grant, which often is every second or third year, is very time consuming. It is valuable in between to consider what you plan to do, but as it is impossible to foresee how your research will develop, you know that half a year or a year later you will probably not be doing what you wrote in your application. It is not only time-consuming to write the applications, but also to evaluate them, meaning that highly qualified scientists spend a lot of time on evaluation, time which could be better used on research.

Second, you never know if you next time will receive support. That is why it is important to be able to present results at the next application, which may tempt to select problems you know can give results. It can be a hindrance for new thinking and for testing new ideas, which may or may not give useful results, but which is so important for the development of basic research. Renewal of the grant also increases the pressure for publication, which may be too early.

Third, the funding system favors successful research. This, of course, deserves support. With 50 to 55 percent of the applications worth supporting, and with money only for 15 to 20 percent, as it is in our country, this leaves little room for new thinking. It is

---

[2] "The Nobel Prize in Chemistry 1997," *Nobelprize.org*. Nobel Media AB 2013, http://www.nobelprize.org/nobel_prizes/chemistry/laureates/1997/.
[3] *Candid Science V*, 448.

often the young who get the new ideas that move borders. It is very difficult for them with few or no previous publications to obtain money to work on their own ideas. Instead, they must join teams who have money and work on the ideas of the head of the team. It is a hindrance for free research. Good research requires quality but also originality and engagement. It stimulates the engagement to work on your own problem and be the author of the paper rather than to work on the problems of the head of the group and be one of many authors on the publication. With the present funding system, it is important that head of groups let the ideas of the single members develop and let them work independently on them, and not just focus on their own ideas.

Fourth, there is the tendency of the politicians to use the funding system to direct the research. The money is put in subject-earmarked boxes. In basic research, it must not be the earmarked subjects which determine the research, but the ideas of the scientists. To let the money determine the research subjects leads to lost possibilities or, in the worst case, to mediocrity.

This does not mean that we shall not have funds. It is necessary for covering the need for grants for expensive chemicals or equipment. But the funds ought to be a supplement to basic amounts of money to departments sufficient for the daily expenses, so the single scientist in a department independently can choose the research subject and work on it without having first to find a grant."[4]

---

[4] Ibid., 451–52

GUNTHER S. STENT

Of the many remarkable people I met during my career, none seemed more secure in his person than Max [Delbrück].

Gunther S. Stent in 2003 in Budapest. (photo by I. Hargittai)

Gunther S. Stent[1] (1924–2008) was born in Berlin and emigrated to the United States at the age of fourteen. He received his PhD degree from the University of Illinois at Urbana-Champaign in 1948. He did postdoctoral work at the California Institute of Technology, the University of Copenhagen, and the Institut Pasteur in Paris. Stent was one of the best-known molecular biologists of his time. He was a prolific author. He published scientific and philosophical books by well-established publishers and self-published his autobiographical memoir *Women, Nazis, and Molecular Biology*. We recorded our conversation in 2003 in Budapest.

[1] B. Hargittai and I. Hargittai, *Candid Science V*, 480–527.

Standing row: Niels Bohr (*far left*), Gunther S. Stent (*fifth from the right*), James D. Watson (*third from the right*), and Herman Kalckar (between Stent and Watson). Sitting row: Élie Wollman, on the right. (courtesy of G. S. Stent)

We covered many topics and concluded by asking Stent for his big-picture view of what the coming years might bring in biology.

"In prognosticating the future of the biological sciences in the post-molecular-biology era, when molecular biology has disappeared as an identifiable specialized discipline, one should bear in mind a deep general principle of the history of science. According to that deep principle, the easier scientific problems are generally solved before the more difficult ones. Thus four difficult, long unsolved problems that had been of central concern for biologists ever since Aristotle founded their discipline in the fourth century BCE were finally worked out during the recently ended twentieth century CE. They are *metabolism, heredity, embryonic development, and organic evolution*.

All four of these ancient problems owe their definitive resolution to the pioneering application of molecular biological ideas and techniques. But in line with the deep historical principle, we can expect that the leftover biological problems that still await their solution in the coming twenty-first century are likely to be more difficult than any that have been solved thus far.

One very difficult unsolved problem is the origin of living matter, which still lacks a credible, coherent proposal for its solution. Perhaps, being a historically unique event that left no traces, the origin of life may be intrinsically insoluble. Despite its obvious importance (and the certainty of the award of a Nobel Prize for its solution), few biologists seem to be working on the origin of life, most likely because of its apparently hopeless intractability. In any case, if it *is* ever solved, molecular biological principles are bound to have played a key role in its solution.

Probably the deepest, as yet unsolved, biological problem, and hence likely to be one of the very last to be resolved, is consciousness. In fact, until recently, the problem of consciousness appeared so deep that it seemed to be a philosophical rather than biological problem."[2]

---

[2] Ibid., 519; see also 519–27 for more on Stent's evaluation of future trends.

JOHN E. SULSTON

Apparently, people achieve things when they're dissatisfied.

John E. Sulston (on the left) and Robert Waterston in 2003 in Cambridge, UK. (photo by I. Hargittai)

John E. Sulston[1] (1942–) was born in Bucks, England. He earned both a bachelor of arts degree in organic chemistry, in 1963, and his PhD degree in oligonucleotide synthesis, in 1966, from the University of Cambridge, UK. From 1966 to 1969, he did postdoctoral research in prebiotic chemistry at the Salk Institute in San Diego, California,. He was affiliated with the MRC Laboratory of Molecular Biology between 1969 and 1992 and has been with what is now the Wellcome Trust Sanger Institute since 1992. He was co-recipient of the 2002

---

[1] B. Hargittai and I. Hargittai, *Candid Science V*, 528–49.

Nobel Prize in Physiology or Medicine with Sydney Brenner and H. Robert Horvitz "for their discoveries concerning genetic regulation of organ development and programmed cell death."[2] We recorded our conversation in 2003 in Sulston's home in Cambridge.

To commemorate the fiftieth anniversary of the double helix model of DNA, there was a meeting in Cambridge, and Sulston gave a passionate talk about the importance of making the human genome information freely accessible to researchers.

"I think I am pragmatic about it. I have become, shall I say, *definite* about it, although you could call it passionate if you like; but I would call it more of a definite position than a passionate position, because it seems to me so simple. It is essential that human genome information is in the public domain. The reason for speaking, if you like, passionately about it is so people understand what the issues are. They think that it's just another batch of data; but we have to consider the principle of how scientific data are handled. This is also why I allowed myself briefly to criticize *Science* magazine in passing, because they actually aided the idea of withholding information from a private company, while at the same time giving them the right to publish their paper in *Science* magazine. To me, it is the antithesis of scientific publication because, in the case of sequence papers, the sequence is the data; it's not something extra which you can take or leave. The extent of this disagreement indicated to me that I must talk about it.

The line is blurred between private research by drug companies and university research.

So I have to explain more about the connection to the human genome. If it had been the case that a private company was simply sequencing the human genome, or any other genome, as those companies do for their own purposes, keeping it private and that's the end of it, that's fine. Then the public domain continues to do those important genomes (which are universally valuable) in its own way, and this has happened many times. I just give you one example. *Staphylococcus aureus*, I gather, has been sequenced ten or maybe twenty times by different companies, because it is important especially for overcoming resistance of that organism to drugs. That's fine, and I have nothing against it, and I would never say that those companies' sequencing should be released. The difference with Celera [Craig Venter's company] was that it deliberately positioned itself as being instead of the public domain and issued misleading statements about the extent to which they would release their data. They claimed to the US Congress that they would release their data every quarter, for example. In fact, many American commentators believed that's actually what happened. You can go back to Congressional Records, to where promises were made to release the data every three months, but it never happened. I'm obviously summarizing all sorts of things that are described in *The Common Thread* [Sulston's book with Georgina Ferry]. They also went and lobbied Congress to try to have public sequencing of the human genome shut down. They told Congress that NIH funds should not be used for such a project.

---

[2] "The Nobel Prize in Physiology or Medicine 2002," *Nobelprize.org*. Nobel Media AB 2013, http://www.nobelprize.org/nobel_prizes/medicine/laureates/2002/.

So this is different, because they are not just doing something on the side that does not effect us all. They are saying that it's appropriate for the human genome to...be in the hands of an American corporation, and for everybody who wants access to it to have to pay; and those who can't pay, of course, can't use it. As I've pointed out, there is an additional disadvantage, had that all happened, even the people who do pay couldn't communicate with each other and publish freely.

Does this happen with any other medical research?

You are quite right about the blurred line. It happened again with rice, for example. There was an American company that was the only one sequencing rice; but, fortunately, their semi-released publication was duplicated by the Chinese. The Chinese did put out their data at the same time and made them available freely. Concerning medical information, I gave examples during the talk, quoting Myriad [Myriad Genetics], for example, but I didn't quote the fee Myriad charged for the sequence of the breast cancer gene, which was USD 2,500. That's an example in medical research, in clinical practice, of a company...using a monopoly which was gained by establishing a patent portfolio—in this case, using a lot of public data because the discovery of the breast cancer genes was very largely done in the public domain. Myriad just managed to establish a few little bits of the sequence. They established a US portfolio and they managed to shut down all other commercial sequencing or any sort of testing of those genes in America. So American physicians who want to have that test have to pay USD 2,500 to Myriad, which is probably about ten times what's necessary. They are also trying to extend that all over the world. They've managed to get the Europeans, the EU, unfortunately, to issue one patent, which is now in litigation, and they are also trying to do this in Toronto, to extend their patent to Canada."[3]

---

[3] *Candid Science V*, 530–32.

HAROLD E. VARMUS

What I've been good at is looking at policy issues from the perspective of a scientist.

Harold E. Varmus in 2002 at the Memorial Sloan-Kettering Cancer Center in New York City. (photo by I. Hargittai)

Harold E. Varmus[1] (1939–) was born in Oceanside on Long Island, New York. He studied English literature at Amherst College and received his doctor of medicine degree from Columbia University in 1966. His research career was at the University of California, San Francisco, which he joined in 1970. Varmus and Michael Bishop shared the 1989 Nobel Prize in Physiology or Medicine "for their discovery of the cellular origin of retroviral oncogenes."[2] Subsequent to his Nobel award, Varmus held several high-level administrative positions. From 1993 to 1999, he was director of the National Institutes of Health; from 2000 to 2010, he was Director of the Memorial Sloan-Kettering Cancer Center,

---

[1] I. Hargittai and M. Hargittai, *Candid Science VI*, 200–215.
[2] "The Nobel Prize in Physiology or Medicine 1970," *Nobelprize.org*. Nobel Media AB 2013, http://www.nobelprize.org/nobel_prizes/medicine/laureates/1970/.

and since 2010, he has directed the National Cancer Institute. We recorded our conversation in 2002 in his office at Sloan-Kettering.

We mentioned that he went into administration soon after his Nobel Prize.

"I did, but I don't like to think of it as administration. I've never been a particularly good administrator. What I've been good at is looking at policy issues from the perspective of a scientist, being reasonable about the way I've pursued things. I've been fairly effective politically, and I don't mean it as an administrator, but politically in both building institutions and getting them supported. I always knew I had the ability to do some of these things, but as a faculty member at the University of California San Francisco, I never took any interest in the kinds of things I would have been asked to do. I've never been interested in being a department chairman and being sure that everybody was getting the right salary and that everyone was having the right amount of space. These are questions that don't interest me. When I look back on what happened to me, I see that I've always been able to run my life the way I wanted to. When I was a faculty member, I was teaching students and did a lot of research, being able to be as productive as I could be. When I had this crown put on my head, I was able to have more influence than I would have had before, and I was able to take on the issues that I really think are important. They had to do with funding of research, training of investigators, support young investigators, reasonable cost policies between government and academia, issues that concern publishing, issues that concern regulations, some of the issues that influence the sharing of research facilities among people, relationships between academia and industry, and misconduct in science. All these are issues that have been much more contentious and much more interesting and reflect on how science is done. These are things I was able to take on once I had this platform for speaking.

Had it not been for the Nobel Prize, you would have stayed, then, a professor?

Sure. I was happy; I was in paradise for a scientist—teaching, doing research, living in a beautiful place, and having a lot of smart friends. It was only the prospect of actually going in one jump that attracted me. If they'd offered me to be the deputy director of some institute, I would've never done that, but to be the head, that was tempting."[3]

Varmus's mother died of breast cancer. We wondered whether this had played any role in his choices of research and administrative positions.

"I assume it played some role in my initially deciding what I was going to work on. She was sick at the time I was making career choices. But there is always the temptation to personalize these choices; there may be some role for it but you also have to see the scientific opportunity. There is no doubt that I learned in my studies that there were things that had not been very well pursued and that could be developed more deeply to better understand cancer. In particular, cancer viruses were sitting there not fully explored, very simple genetic units that made normal cells behave like cancer cells. It

---

[3] *Candid Science VI*, 204.

was clear that there was a revolution about to occur in these studies as a consequence of molecular biology. Even though it was still not possible to clone DNA or sequence DNA, it was possible to try to understand how single genes of the kind that we found in tumor viruses worked, where those genes came from, how they behaved in various circumstances. So the temptation to enter this field was very strong, even without a familial connection. We all know that cancer is a major health problem in our society, and everybody knows people among their friends and relatives who had cancer."[4]

Being a top administrator, Varmus may have a broad overview from which to prognosticate about the thrust of science in the coming years.

"It will still be molecular biology. Molecular biology has changed somewhat, in large part by sequencing genomes. It will also be changed by the impact of computational science and databases. Imaging is also changing biology, the ability to see molecules, to see how cells operate, and to follow their fate. There's an increasing tendency for biologists to think visually. The three-dimensional picture of how cells are working is important. There are physical tools. Then there is an enhanced role for chemistry. Chemical biology is one of our slogans. Chemistry may have a bad image in the eyes of the general public, but in our science, exploring chemical space, building new substances with combinatorial chemistry are of great interest and therapeutically promising. The merger of structural biology—that is, the three-dimensional structure of proteins—with potential drugs, is among the most important themes in medical science."[5]

---

[4] Ibid., 207.
[5] Ibid., 215.

ALEXANDER VARSHAVSKY

Doing science is like driving a car at night. You can only see as far as your headlights, but you can make the whole trip that way.

Alexander Varshavsky in 2004 at Caltech. (photo by I. Hargittai)

Alexander Varshavsky[1] (1946–) was born in Moscow. He received his master's degree equivalent from the Faculty of Chemistry, Moscow State University, and his PhD equivalent degree from the Institute of Molecular Biology of the Soviet Academy of Sciences. In 1977, he defected to the West. First, he was at the Massachusetts Institute of Technology, and then moved to Caltech, where he is currently Smits Professor of Cell Biology. His main research interest has been the biological aspects of the ubiquitin systems, and he has made numerous discoveries. He has earned a number of the most prestigious prizes, including the Wolf Prize in Medicine (2001) and the Lasker Award in Basic Medical Research (2000). The latter he shared with the 2003 chemistry Nobel laureates

---

[1] I. Hargittai and M. Hargittai, *Candid Science VI*, 310–59.

Avram Hershko and Aaron Ciechanover. We had preliminary conversations in 1999 and 2004, and did a recording in 2005 in Pasadena, California.

On doing science.

"Doing science is like driving a car at night. You can only see as far as your headlights, but you can make the whole trip that way. Long before I encountered that metaphor, by E. L. Doctorow (his actual remark was about writing a novel), I sensed the attraction of the scientist racket: an air of open-ended adventure, a contest of sorts, with the landscape rough, unpredictable, with other cars racing toward Holy Grails out there, and occasionally colliding with yours, by accident or not quite. A rambunctious life, to be sure, but quiet on the outside, with maelstroms and lava flows hidden from casual view. A set of qualities for such a life must contain a genuine interest, beyond mere curiosity, in understanding the world's design, but an ambition too, even with thinkers whose visage suggests otherwise. Photos of old Einstein, a serene photogenic sage, free of strife, do not recall the assertive and ambitious young man in Switzerland, on the cusp of initial success, and his later, often overt competition with rivals, the great German mathematician David Hilbert amongst them. Hilbert produced (and correctly interpreted) the equations of general relativity simultaneously with Einstein, a fact unmentioned in popular accounts of the subject but known to those who are interested in physics' history. *What Mad Pursuit* is the title, from a poem by Keats that Francis Crick chose for his autobiography. And mad it is, propelled not only by desire for knowledge but also, in no small measure, by desire to impress and awe—oneself, others, posterity. That's where one's genetic makeup comes in particularly strongly, I think: a predilection for life of a certain shape and texture.

My mother tells me that her ancestry is traceable to the chief rabbi of Prague in the early seventeenth century. Had I been born a few centuries earlier in the Jewish ghetto of Prague, I might have ended up a religious scholar, splitting hairs with fellows at a yeshiva about the subtleties of the Talmud and Kabala. It's nice to hope that I would have discerned, unaided, the incorrectness of religious outlook, recognizing its vacuity in the midst of semi-medieval Prague. Wrong hope, most likely, for it is difficult, nowadays, to appreciate the acuteness of insight. In addition to the independence of spirit that would be necessary for such a discovery before the rise of modern science. But some, quite rare, people in both antiquity and the Middle Ages must have glimpsed this insight. That the names of early doubters are largely unknown to us is no accident. From time immemorial to roughly the 1700s in Europe, one could safely declare a disbelief in deities and their deeds only in total solitude. Even centuries later, the nonbeliever's view of extant religions is a majority opinion only among scientists, not in the public at large. But the long-term trend, on the scale of centuries, although a turbulent one, with transient local reverses, is clearly toward a secular, science-informed outlook. A stance in which things and phenomena we don't have good explanations for are called mysteries or puzzles, without attempts to camouflage the insufficient understanding by theology and fairy tales."[2]

---

[2] Ibid., 320–21.

About science in general.

"'He digs deep, but not where it's buried,' averred the poet Anna Akhmatova, in a conversation cited by her biographer. She was speaking, naturally, of another poet. (Compared to writers, scientists positively love each other.) Akhmatova's unkind aphorism captures the predicament of anyone who aspires to innovation, be it a momentous way of stringing words together or major discoveries in science. There is a distinction, here, between a poet and scientist. A poet may despair of finding a way to connect an insight and its form of expression. He may never find that form, or he may find it the next minute. If he does, the result, a verse, is truly his own. While poetry is occasionally about content, it is primarily an alloy of content and form, and the form is poet-specific. (Hence Robert Frost's definition of poetry as the part that gets 'lost in translation'.) So the poet is, in a way, safe from being scooped. Other dangers, aplenty. But not that one. By contrast, in science a truth is a stickler for accuracy but cares little about the form. One must arrive at a truth first and mark the arrival by a published paper. The form, that unique identifier of the individual, barely counts in science, and certainly does not in the long run. Hence the extreme competitiveness of scientists throughout the ages. Their genuine curiosity about the world is warped and smothered by their haste to acquire narrow expertise and tools, lest that buried fruit is unearthed by another digger who is simply lucky, or better prepared, or (most often) both. The only thing I dislike about working in my beloved profession is my inability to enjoy a study by learning at a leisurely, pleasant pace, as broadly and gradually as I care to, instead of being focused and intense. There must be scientists of sunny, relaxed, unhurried dispositions, but I never met such people amongst the peers whose discoveries are both first-rate and...number more than one. Perhaps they are tranquil when they retire or become administrators. But they are not mellow in their prime."[3]

---

[3] Ibid., 357–58.

JAMES D. WATSON

Any good chemist should've found the structure of DNA. But The Double Helix *was probably unlikely to have been written by anyone beside myself.*

James D. Watson with Elizabeth (Liz) Watson (on the left) and Magdolna Hargittai in 2000 in Budapest. (photo by I. Hargittai)

James D. Watson[1] (1928–) was born in Chicago. He received his undergraduate degree from the University of Chicago and his PhD degree from Indiana University, where Salvador Luria was his mentor. He did postdoctoral work in Denmark. Then came his stay in Cambridge, UK, and the iconic discovery of the double helix together with Francis Crick, for which Watson and Crick, together with Maurice Wilkins, were awarded the 1962 Nobel Prize in Physiology or Medicine. He is also famous for his books, especially for *The Double Helix*, an innovative and highly personal narrative of their seminal discovery. His other achievements include the development of the Cold Spring

[1] I. Hargittai and M. Hargittai, *Candid Science II*, 2–15.

Harbor Laboratory on Long Island, New York, into an excellent research center and graduate school. We recorded three conversations with him in his office at the Cold Spring Harbor Laboratory; the first in 2000 and the other two in 2002. *Candid Science II* contained parts of the first conversation while the book *The DNA Doctor*[2] gives meticulous account of all three.

On his longest-ranging impact.

"Probably my books. The discovery of DNA was just waiting to be made; it was not a difficult thing; any good chemist should've got the answer pretty fast. Rosalind Franklin was a physical chemist; she really wasn't a complete chemist. Pauling just goofed beyond any reason; it was crazy. Any good chemist should've found the structure of DNA. But *The Double Helix* was probably unlikely to have been written by anyone beside myself.[3]

Intelligence and genetics—a taboo question?

For some people, yes. It's difficult to define intelligence while we don't really know how the brain works. I've always thought I have no mathematical ability, and other people would say you have no mathematical interest and therefore you were not motivated to learn it. It's very hard to distinguish them. I think my brain works fast on what interests me. We don't know how we store numbers in the brain; we don't know any of these things.

But you are saying, "The misuse of genetics by Hitler should not deny its use today." What bothers me (IH) in this is that any demagogue says a lot of things that have a grain of truth in it yet the total is totally false. Couldn't we approach this whole problem [of the use of genetics] without invoking Hitler and Nazism?

No, because if I go out in the public, they say, 'You're playing Hitler.' If other people didn't raise it, I wouldn't, but they constantly want to talk about the 'eugenic past.' The German disdain of the fact that the Greens have been so effective in preventing DNA-based industry from developing, at least they were in Germany, had a lot to do with the fact that genetics had become identified with Hitler. But genetics had been identified with Stalin too. Hitler used genetics arguments; Stalin denied its existence. You had two extremes. You have to know what's right and what's wrong. Hitler wanted the German race to be perfect, so he wanted to kill off the imperfect Germans while they were young.[4]

Genetically modified food? You may have seen the European resistance to it.

They're trying to bring it over here [to the US]. I'll give you Prince Charles's latest [anti-GMF] statement. You can have fun with that. They've asked me to reply to it because the British people don't find it easy to attack him. He's totally misguided; but he dislikes modern architecture, too. He's caught in the past and would like to go

---

[2] Istvan Hargittai, *The DNA Doctor: Candid Conversations with James D. Watson* (Singapore: World Scientific, 2007).
[3] Ibid., 37.
[4] Ibid., 47–48.

back to when kings were important. He's harmful to the people. If Charles were king, I would be a republican."[5]

On the utilization of genetics.

"Right now it's not the misuse of genetics, rather the disuse of it, that we're not using it. We're not screening people for cystic fibrosis.
  Why not?
  There are two reasons. People are afraid to look into the future, and they don't want to discover that maybe they're carrying something that could cause them harm in the future. That's partly it and partly, essentially, the Right to Life movement doesn't want genetic knowledge to be used; tests, which might lead to abortion, are thought to be inherently bad.... The true miracles in the world come from science, new ideas, the curing of disease, smallpox. Basically, the scientists are the people who do the impossible. The only people who do the impossible are those who save people, and that comes from knowledge. It doesn't come from prayer. The important thing is knowledge, not prayer.[6]
  A message in this connection?
  Because of science, people have better lives today than before, and in 2010 life will be yet better. But I don't think too far ahead. We are where we are because we use knowledge and we have culture, which we can store now. There will always be uncertainty. The thing that finally could do us in would be some form of disease that we're not prepared for.
  The main thing is you don't kill someone because you dislike him. I think genetics will help people more but looking into the future is scary. I don't want to know when I go to the doctor what he'll say. But, on the other hand, you may look into your future and say that your blood pressure is high and we can lower it now and you don't have to die of high blood pressure the way you did fifty years ago."[7]
  MIT professor Nancy Hopkins, Watson's former student, stated that Watson was the first feminist she ever met. So we wondered whether or not Watson considered himself a feminist.
  "No, not by someone's standards. I've been a strong supporter of women who intend to do science. I treated them like anyone else, maybe a little better because of sex discrimination. I don't know the definition of feminist and that's why I'm avoiding the term. But I don't know of any case when we showed prejudice in not accepting a woman applicant; nor do I know of any case when we showed any prejudice in dealing with them as far as exams or things like that are concerned. At Harvard, our department offered a faculty position to a woman. That was very early on, some time in the mid-sixties.
  Do you think that women are taken as seriously as men in science?
  I don't really want to answer that because I'm not in a department... So I can't say anything about it, only to the extent of my knowledge. You'd have to ask others. I think

---

[5] Ibid., 72.
[6] Ibid., 77, 79.
[7] Ibid., 80.

the situation is inherently made awkward by sexual harassment charges. In the case of men, you can assume that you cannot promote someone and that's the decision. In case of a woman, even though you think she is incompetent, you feel pretty sure that in many cases lawyers will appear and accuse you of harassment. That's what makes you reluctant to hire women because you have the feeling that you can't essentially fire them.

Does it mean that positive discrimination would not be good for women?

It makes me reluctant because I can be fairly honest and tough to a man, but if I feel that way to a woman, it is bad, though not as bad as with blacks, but it is half way there."[8]

---

[8] Ibid., 140–42.

CHARLES WEISSMANN

There may be a circumstance where there would be a good medical and humanitarian reason to do it [clone humans].

Julie and Charles Weissmann in 2000 in London. (photo by I. Hargittai)

Charles Weissmann[1] (1931–) is Professor Emeritus of the Department of Infectious Diseases at the Scripps Research Institute in Jupiter, Florida. He is Swiss, but was born in Hungary when his mother was in Budapest visiting a friend. Mother and child returned to Switzerland a few weeks later. Weissmann received his schooling in Zurich and earned his doctor of medicine degree in 1956 and his PhD degree in 1961 at the University of Zurich. Following a few years at New York University (1961–1967), he spent the major part of his career at the University of Zurich. After retiring, he was affiliated with

---

[1] I. Hargittai and M. Hargittai, *Candid Science II*, 466–97.

the MRC Neurogenetics Unit in London, and lately, he has been with Scripps Florida. He is well-known for his work on RNA phage replication and the gene expression of eukaryotic cells, reverse genetics, retroviruses, interferon, and protein cloning, including the proteins involved in scrapie and prion diseases. We recorded our conversation in 2000 in London.

At some point, some scientists found it hard to accept that (prion) proteins may transmit diseases, just as a few decades earlier the idea that DNA was the genetic material had a difficult time gaining acceptance. Weissmann spoke to us about this.

"You are exactly right. Then, most everybody believed that protein was the genetic material and DNA was supposed to be some junk—a storage form of nucleotides was one proposal. Avery showed the genetic material was DNA. In the early 1980s, everybody believed the scrapie agent was some unusual virus, and we were claiming it was a protein.

At that point, many scientists who were not directly involved in this research, who were not emotionally compromised, were becoming convinced. Many researchers, however, who had spent most of their scientific lives looking for a virus, found it difficult to accept that they had been looking for something that doesn't exist. There are a few of the old guard who are still reluctant to accept the protein-only hypothesis.

In the late 1980s, it occurred to me that if we could eliminate the PrP from the mouse, this mouse should become resistant to the disease and incapable of replicating the agent. When I learnt from Peter Gruss that he and others were beginning to develop methods to destroy, in specific, targeted fashion, individual genes in a mouse, I decided we should follow in their footsteps. One of my very talented collaborators, Michel Aguet, undertook to develop this method to knock out the PrP gene. Mario Capecchi and Oliver Smithies in the United States were successful in carrying out targeted 'gene knockouts' and we learned essentially how to do these things from their work. Michel Aguet and a graduate student, Hansruedi Büeler, carried out our experiment. It was a tedious, long, and difficult procedure at the time. Finally, we generated these mice and we were astonished by what we hardly had dared to hope for, namely, that animals devoid of PrP developed normally and were healthy. There were no obvious defects to these animals. This was very surprising because here was a gene that was conserved during evolution in all vertebrates, was expressed mainly in the brain and in many other organs, already early in embryonic development, and yet it could be eliminated without causing damage. Why this is so is a chapter for itself. We were, of course, very happy because now we could do our experiments. The reason somebody said to you that maybe I should have shared the Nobel Prize, may have been because that was probably the most convincing experiment in favor of the protein-only hypothesis. One must be very clear though that what this experiment proves is that PrP is essential for generating the infectious agent; it doesn't prove that it *is* the infectious agent, although this is by far the simplest interpretation.

I'm a rather careful person; I worry about making mistakes and we were very careful to point out what the experiment really meant. However, most people said that this proved that the protein is the infectious agent. It was certainly a good argument.

To summarize, we cloned the gene and we did this knockout experiment. Stan's [Prusiner] merit was that he isolated this protein, which led us to PrP and to the gene; he showed the linkage between familial forms of the disease and the gene, and he showed the gene could facilitate crossing the species barrier. The additional merit of Stan is that he was very passionate about this. He was convinced that he was right even when nobody would listen to him, and he was subjected to a lot of hostility and vituperation. Some of the early work and early papers were not very convincing. Many of his experiments were not very clean and he tended to over-interpret them. This drew quite a lot of criticism. But the thrust of everything was correct. If you are looking at the relative merits, Stan was the person who carried the ball when nobody was willing to listen to him. I came in skeptical, but through our own experiences, I became convinced that the protein-only hypothesis is correct. Our experiment with the gene knockout was the kind of experiment one does to falsify a hypothesis, and when this experiment spectacularly failed to do so, I was also convinced.

Stan was very intent on getting the Nobel Prize and he really worked hard at it. He asked people to propose him, including myself, which I did. As for myself, I felt that if there was merit in what I've done, it would be recognized without my having to do anything about it. And I did receive a number of other awards.

What is the relationship between the prion work and Gajdusek's work on kuru?

Gajdusek is a brilliant person, a polymath. He was sent to Papua New Guinea to investigate kuru. At that time people didn't know whether kuru was a result of poisoning by some plant or animal these natives were eating or whether it was a nutritional deficiency or whether it was genetic disease. Almost nobody thought it was transmissible. Gajdusek's big merit is that he showed that it is transmissible. He had done several transmission experiments and when nothing happened after a few weeks, the experiments were discarded. Usually, when something is transmissible, you see the disease after a few weeks, whether it's a virus or bacteria. Gajdusek collected a number of brain specimens and he sent them back to the States; they made the sections and they found the spongiform degeneration and the deposition of an amyloid in the form of what is called kuru plaques. Stan later showed that they consisted mainly of PrP^{Sc}. Those pictures then were exhibited in London and a man called [William J.] Hadlow, who was a scrapie researcher, saw them and noticed that they looked exactly like scrapie. He published a short note in *Lancet*, pointing out the similarity between scrapie and kuru and he also wrote a letter to Gajdusek. At that time it was already known that the incubation time of scrapie was extremely long, years or decades. The suggestion was that if kuru was like scrapie, it wouldn't be enough to wait a few weeks, you would have to wait a few years. Gajdusek and his colleagues followed up on this and after two years the inoculated chimpanzees got sick. Stan always points out that Hadlow should have shared the Nobel Prize with Gajdusek, because if Hadlow had not made his remark, Gajdusek would likely not have done the long-term experiment."[2]

---

[2] Ibid., 472–474.

ROSALYN S. YALOW

We were very determined not to take out a patent on it [RIA] but to do everything to help it spread.

Rosalyn S. Yalow with Solomon Berson. (courtesy of R. S. Yalow)

Rosalyn Yalow,[1] née Sussman (1921–2011), was born in New York City. She received her bachelor's degree in physics and chemistry from Hunter College in 1941 and her master of science and PhD degrees in nuclear physics from the University of Illinois at Urbana in 1942 and 1945, respectively. She worked almost from the beginning of her career at the Bronx Veterans Administration Medical Center. For many years, the physicist Yalow worked together with the medical doctor Solomon Berson. They made seminal discoveries of which

[1] I. Hargittai and M. Hargittai, *Candid Science II*, 518–23.

their new technique called radioimmunoassay (RIA) became the most famous and widely used. RIA allows for the determination of minute amounts of hormones and other vital substances in the blood.

It was expected that Berson and Yalow would one day receive the highest recognition for their discoveries, but then Berson died. Yalow's hopes might have been shattered, but she doubled her efforts, not so much to win prize, but to prove herself that she was not merely Berson's assistant. She succeeded and in time the Nobel recognition also arrived. She received it in 1977 in Physiology or Medicine "for the development of radioimmunoassays of peptide hormones."[2]

We recorded our conversation in 1998 in her office at the Bronx Veterans Administration Medical Center. By then, she had suffered a stroke that had left her partially paralyzed.

She told us about her life.

"My mother was four when her family came to this country; and my father was born in New York. They both came from poor East European immigrant families. She completed the sixth grade and he the fourth grade only, so they didn't even have the advantage of a high school education. But they were determined for their children to have a college education.

I went to school in the Bronx. It was not a great school but we had good teachers and the pupils were very motivated. They were predominantly poor and Jewish. I went to Hunter College, which was part of the New York City College system. The boys went to City College and the girls went to Hunter College.

I was very lucky. There were three physics professors at Hunter who took me under their wings, and I couldn't have imagined anything else for myself than becoming a physicist. It wasn't easy, but eventually I was offered a teaching assistantship at the University of Illinois at Urbana. They brought in three Jews at Urbana that year, two boys in addition to myself. One of them was my future husband, Aaron Yalow. My thesis director was Maurice Goldhaber, who was later to become Director of the Brookhaven National Laboratories.

The war made it possible for me and for many other young Jewish students to enter graduate school. While in Europe the Jews were being killed, the war made all the difference for Jews and for women in America.

I did my graduate research in nuclear physics. This was very much the thing to do at that time. After receiving my PhD, I did some teaching at Hunter and, eventually, I became an associate of the Bronx Veterans Administration. This place used to be much larger than it is now. It used to have 1200 beds, and now there are about 500. The war veterans are dying off, and many of the surviving veterans go to small private hospitals.

---

[2] "The Nobel Prize in Physiology or Medicine 1977," *Nobelprize.org*. Nobel Media AB 2013, http://www.nobelprize.org/nobel_prizes/medicine/laureates/1977/.

My husband died five years ago, but we had a wonderful marriage. He was always very supportive. He was Professor of Physics at Cooper Union College here in New York. He was engaged in teaching, not research. We have two children. When they were small, in the 1950s, my mother used to come to help us with the children. We also had maids, first a sleep-in maid, later a part-time maid. They were wonderful, bright, black women from the South. They came to New York, and they couldn't go to study, so they found work as maids. This made it possible for me to carry on with my job. For me, this was a very fortunate situation. Today this would be impossible.

Our daughter, Elanna, studied educational psychology and has her PhD. She lives in San Francisco, has two children, and works all over the country, setting up day-care centers. Our son, Benjamin, used to do computer work. He doesn't have a formal job currently. I live together with him. I've had three strokes. I have difficulties in moving my right hand, and my right leg is partially paralyzed. I come to my office regularly, though, read my mail, keep up with things around me."[3]

Berson and Yalow acted in unison in their work and together they decided to not patent their radioimmunoassay.

"This was a conscious decision on our part. The radioimmunoassay technique was a joint discovery by myself and Solomon A. Berson. Sol and I worked together for eighteen years before he left the Veterans Administration and became Chairman of Medicine at the Mount Sinai School of Medicine in 1968. He died four years later. In 1972, a young doctor of medicine, Eugene Straus joined me.

When Sol and I discovered the radioimmunoassay technique, at the beginning it had a slow start, but we knew it would catch up quickly because it was a very sensitive and very useful tool. We were very determined not to take out a patent on it but to do everything to help it spread. We organized courses to teach physicians to use the technique.

Sol and I made a great team. For many years we shared the same office. We had two desks in the office, and we were discussing things all the time. I never had any formal training in biology. He taught me everything I needed in biology and medicine, and I taught him some physics. He knew a lot of physics, but he was an doctor of medicine. Compared with a university setting, we had the great advantage of having no competition between us."[4]

---

[3] *Candid Science II*, 520–22.
[4] Ibid., 523.

LIST OF NAMES

The asterisks indicate names of interviewees excerpts from whose conversations are presented in this volume.

't Hooft,* Gerard (1946–). Dutch physicist; Nobel laureate (1999)
Abelson, Philip H. (1913–2004). American physicist and science writer
Abrikosov, Alexei, A. (1928–). Soviet-American physicist; Nobel laureate (2003)
Aguet, Michel. Swiss molecular biologist
Akhmatova, Anna (1889–1966). Russian poet
Aleksandrov, Anatoly P. (1903–1994). Soviet physicist and science administrator
Alferov,* Zhores I. (1930–). Soviet-Russian physicist; Nobel laureate (2000)
Alpher, Ralph (1921–2007). American physicist
Anderson,* Philip W. (1923–). American physicist; Nobel laureate (1977)
Ann (1950–). British royal Princess
Arber,* Werner (1929–). Swiss biologist; Nobel laureate (1978)
Archimedes (287–212 BCE). Greek mathematician and scientist
Aristotle (384–322 BCE). Greek philosopher and polymath
Astbury, William T. (1898–1961). British physicist
Avery Oswald T. (1877–1955), American medical researcher
Ax, Emanuel (1949–). Polish-American pianist
Axelrod, Julius (1912–2004). American biochemist; Nobel laureate (1970)
Bacon, Francis (1561–1626). English philosopher
Baltimore,* David (1938–). American biologist and science administrator; Nobel laureate (1975)
Banting, Frederick G. (1891–1941). American biomedical scientist; Nobel laureate (1923)
Bárány, Anders (1942–). Swedish physicist; Secretary of the Nobel Committee for Physics (1989–2003)
Basov, Nikolai G. (1922–2001). Soviet physicist; Nobel laureate (1964)
Bastiansen, Otto (1918–1995). Norwegian chemist and science administrator
Baylen, Sarah (1916–2005). Herbert C. Brown's wife
Beadle, George W. (1903–1989). American geneticist; Nobel laureate (1958)
Bekesy, George (1899–1972). Hungarian-American biophysicist; Nobel laureate (1961)
Bell Burnell,* Jocelyn (1943–). British astrophysicist
Ben Gurion, David (1886–1973). Israeli politician; the first prime minister of Israel (1948–1954; 1955–1963)
Benzer,* Seymour (1921–2007). American physicist turned biologist
Berg,* Paul (1926–). American biomedical scientist; Nobel laureate (1980)
Bernard, Claude (1813–1878). French physiologist
Berson, Solomon (1918–1972). American physician and researcher
Best, Charles (1899–1978). American-Canadian medical scientist; co-discoverer of insulin

Bethe, Hans (1906–2005). German-American physicist; Nobel laureate (1967)
Bishop, J. Michael (1936–). American biomedical scientist; Nobel laureate (1989)
Blaschko, Hermann (1900–1993). German-British biomedical scientist
Blumberg,* Baruch S. (1925–2011). American biomedical scientist; Nobel laureate (1976)
Bohr, Niels (1885–1962). Danish physicist; Nobel laureate (1922)
Bragg, W. Lawrence (1890–1971). British physicist; Nobel laureate (1915)
Brahe, Tycho (1546–1901). Danish astronomer
Bréchignac, Philippe. French physicist, Catherine Bréchignac's husband
Bréchignac,* Catherine (1946–). French physicist and science administrator
Breit, Gregory (1899–1981). Russian-American physicist
Brenner,* Sydney (1927–). British biologist; Nobel laureate (2002)
Brockway, Lawrence (1907–1979). American chemist
Brodie, Bernard B. (1909–1989). American pharmacologist
Brown,* Herbert C. (1912–2004). American chemist; Nobel laureate (1979)
Brownlee, George. British chemical pathologist
Büeler, Hansruedi (1963–). Swiss-American biomedical scientist
Burnet, Frank Macfarlane (1899–1985). Australian virologist; Nobel laureate (1960)
Burroughs, Silas M. (1846–1895). American pharmacist, businessman, and philanthropist
Cabot Lodge, Henry, Jr. (1902–1985). American politician
Cairns, John F. (1922–). British biomedical scientist
Calne, Roy Y. (1930–). British surgeon and pioneer in organ transplantation
Capecchi, Mario (1937–). Italian-American biologist; Nobel laureate (2007)
Carlsson,* Arvid (1923–). Swedish biomedical scientist; Nobel laureate (2000)
Carter, Jimmy (1930–). 39th US president (1977–1981)
Chadwick, James (1891–1974). British physicist; Nobel laureate (1935)
Chain, Boris (1906–1979). German-British biochemist; Nobel laureate (1945)
Chandrasekhar, Subrahmanyan (1910–1995). Indian-American astrophysicist; Nobel laureate (1983)
Chargaff,* Erwin (1905–2002). Austrian-American biochemist
Charlemagne (742–814). Holy Roman Emperor (774–814)
Charles (1948–). Prince of Wales
Cheronis, Nicholas D. (1896–1962). American chemist
Chiang, Kai–shek (1887–1975). Chinese political leader
Chibnall, Albert C. (1894–1988). British biochemist
Chirkov Nikolai M. (1908–1972). Soviet chemist
Churchill, Winston (1874–1965). British politician, author; Nobel laureate (1953)
Ciechanover,* Aaron (1947–). Israeli biomedical scientist; Nobel laureate (2004)
Clemett, Simon (1968–). American chemist
Clinton, William (Bill) J. (1946–). 42nd US President (1993–2001)
Cohen, Stanley (1922–). American biochemist; Nobel laureate (1986)
Cohn,* Mildred (1913–2009). American chemist
Columbus, Christopher (1451–1506). Italian explorer
Conant, James B. (1893–1978). American chemist and civil servant
Conway,* John H. (1937–). English-American mathematician
Cooper, Leon N (1930–). American physicist; Nobel laureate (1972)
Copernicus, Nicolaus (1473–1543). Polish astronomer and polymath
Cormack, Alan M. (1924–1998). South African-American physicist; Nobel laureate (1979)

359 LIST OF NAMES

Cornell, Eric A. (1961–). American physicist; Nobel laureate (2001)
Cornforth, Rita H., née Harradence; British chemist
Cornforth,* John W. (1917–). British chemist; Nobel laureate (1975)
Cotton, F. Albert (1930–2007). American chemist
Cotzias, George (1918–1977). Greek-American biomedical scientist
Cram,* Donald J. (1919–2001). American chemist; Nobel laureate (1987)
Crick, Odile, née Speed (1920–2007). British artist; Francis Crick's wife
Crick,* Francis (1916–2004). British physicist turned biologist; Nobel laureate (1962)
Cronin, Alexander (1970–). American physicist
Crutzen,* Paul J. (1933–). Dutch scientist; Nobel laureate (1995)
Curl, Robert F. (1933–). American chemist; Nobel laureate (1996)
Dalton, John (1766–1844). English chemist
Darwin, Charles (1809–1882). English naturalist
Davy, Humphry (1778–1829). British chemist
Dawson, Joan M.. American biomedical scientist; Paul Lauterbur's wife
Degkwitz, R. (1920–1990). German biochemist
Deguchi, Takeo. Japanese biochemist
Deisenhofer,* Johann (1943–). German biochemist; Nobel laureate
Delbrück, Manny (1917–1998). Max Delbrück's wife
Delbrück, Max (1906–1981). German-American physicist turned biologist; Nobel laureate (1969)
Dirac, Paul A. M. (1902–1984). British physicist; Nobel laureate (1933)
Djerassi,* Carl (1923–). American chemist, author, playwright
Doctorow, E. L. (1931–). American author
Dresselhaus, Eugene (1929–). American physicist; Mildred S. Dresselhaus's husband
Dresselhaus,* Mildred S., née Spiwak (1930–). American physicist
Drucker, Peter (1909–2005). Austrian-American educator and author
Dulbecco, Renato (1914–2012). Italian-American biomedical scientist; Nobel laureate (1975)
Dyson,* Freeman J. (1923–). British-American physicist and author
Eigen, Manfred (1927–). German chemist; Nobel laureate (1967)
Einstein, Albert (1879–1955). German-Swiss-American physicist; Nobel laureate (1922 for 1921)
Eisenhower, Dwight D. (1890–1969). American military leader; 34th US president (1953–1961)
Elion,* Gertrude B. (1918–1999). American chemist and pharmacologist; Nobel laureate (1988)
Elizabeth II (1926–). Queen of Great Britain and Northern Ireland (1953–)
Enders, John F. (1897–1985). American biomedical scientist; Nobel laureate (1954)
Ernst, Richard (1933–). Swiss chemist; Nobel laureate (1991)
Eschenmoser,* Albert (1925–). Swiss chemist
Evenson, Robert E. (1935–2013). American economist
Fairlie, David B. (1935–). American mathematician
Faraday, Michael (1791–1867). British scientist
Feinberg, Evgenii L. (1912–2005). Soviet physicist
Fermi, Enrico (1901–1954). Italian-American physicist; Nobel laureate (1938)
Ferry, Georgina (1955–). British science writer
Feynman, Richard (1918–1988). American physicist; Nobel laureate (1965)
Fisher, Michael E. (1931–). English-American physicist
Fleming, Alexander (1881–1945). British biomedical scientist; Nobel laureate (1945

Florey, Howard W. (1898–1968). Australian pharmacologist; Nobel laureate (1945)
Fothergill, Leroy D. (1934–). American biomedical scientist and arms control advisor
Fowler, Ralph H. (1889–1944). British physicist
Fraenkel, George K. (1921–2009). American physical chemist
Franklin, Rosalind (1920–1958). British biophysicist
Frederick II (1534–1588). King of Denmark and Norway (1559–1588)
Freund, Peter (1936–). Romanian-American physicist
Friedman,* Jerome I. (1930–). American physicist; Nobel laureate (1990)
Frost, Robert (1874–1963). American poet
Fukui,* Kenichi (1918–1998). Japanese chemist; Nobel laureate (1981)
Gajdusek,* D. Carleton (1923–2008). American biomedical scientist; Nobel laureate (1976)
Galich, Aleksandr (1918–1977). Russian poet, playwright, dissident.
Galilei, Galileo (1564–1642). Italian scientist
Galpern,* Elena (1935–). Soviet-Russian chemist
Gamow, George (Georgii A.) (1904–1968). Russian-American physicist and author of popular science books
Garwin,* Richard L. (1928–). American physicist
Gell-Mann, Murray (1929–). American physicist; Nobel laureate (1969)
Gilbert,* Walter (1932–). American physicist and biologist; Nobel laureate (1980)
Gilcrease, F. Wellington. Former New York State Health Department official
Ginzburg,* Vitaly L. (1916–2009). Soviet physicist; Nobel laureate (2003)
Glaser,* Donald A. (1926–2013). American physicist turned biologist; Nobel laureate (1960)
Glashow, Sheldon L. (1932–). American physicist; Nobel laureate (1979)
Goeppert Mayer, Maria (1906–1972).). German-American physicist; Nobel laureate (1969)
Goering, Hermann (1893–1946). German Nazi leader
Goldhaber, Gertrude, née Scharff (1911–1998). German-American physicist; Maurice Goldhaber's wife
Goldhaber,* Maurice (1911–2011). Austrian-American physicist
Greengard, Paul (1925–). American biomedical scientist; Nobel laureate (2000)
Greytak, Thomas. American physicist
Gross,* David (1941–). American physicist; Nobel laureate (2004)
Gruss, Peter (1949–). German biologist
Haber, Fritz (1868–1934). German chemist; Nobel laureate (1919, for 1918)
Hadlow, William J. (1921–). British scientist; recognized that human kuru and scrapie in sheep looked alike
Hahn, Otto (1879–1968). German chemist; Nobel laureate (1945 for 1944)
Halford, Ralph S. (1914–1978). American chemist
Hamaroff, Stewart (1947–). American biomedical scientist
Hamburger, Viktor (1900–2001). German-American biologist
Hammett, Louis P. (1894–1987). American chemist
Hartwell, Leland H. (1939–). American biologist; Nobel laureate (2001)
Heidelberger, Michael (1888–1991). American immunologist
Heilbron, Ian (1886–1959). British chemist
Heisenberg, Werner (1901–1976). German physicist; Nobel laureate (1933 for 1932)
Herman, Helen; Robert Herman's wife
Herman, Robert (1914–1997). American physicist
Herschbach, Dudley R. (1932–). American chemist; Nobel laureate (1986)

361 LIST OF NAMES

Hershko, Judy. Avram Hershko's wife
Hershko,* Avram (1937–). Israeli biomedical scientist; Nobel laureate (2004)
Hertz, Gustav L. (1887–1975). German physicist; Nobel laureate (1925)
Hewish,* Antony (1924–). English astrophysicist; Nobel laureate (1974)
Hilbert, David (1862–1943). German mathematician
Hitchings, George H. (1905–1998). American pharmacologist; Nobel laureate (1988)
Hitler Adolf (1889–1945). German Nazi dictator
Hochstrasser, Robin M. (1931–2013). American chemist
Hoffman,* Darleane C. (1926–). American chemist
Hoffmann,* Roald (1937–) American chemist, poet, and author; Nobel laureate (1981)
Hofstadter, Robert (1915–1990). American physicist; Nobel laureate (1961)
Holley, Robert W. (1922–1993). American biochemist; Nobel laureate (1968)
Holmes, Frederick L. (1932–2003). American science historian
Hopkins, Nancy; American molecular biologist
Hornykiewicz,* Oleh (1926–). Austrian biomedical scientist
Horvitz, H. Robert (1947–). American biologist; Nobel laureate (2002)
Hounsfield, Godfrey N. (1919–2004). British engineer; Nobel laureate (1979)
Hoyle, Fred (1915–2001). British astronomer
Hubble, Edwin P. (1889–1953). American astronomer
Huber, Robert (1937–). German biochemist; Nobel laureate (1988)
Huffman, Wallace E.; American agricultural economist
Hulse, Russel A. (1950–). American physicist; Nobel laureate (1993)
Hunt, R. Timothy (1943–). British biochemist; Nobel laureate (2001)
Hutchins, Robert M. (1899–1977). Innovative president of the University of Chicago
Ioffe, Abram F. (1880–1960). Russian-Soviet physicist; mentor of many renowned Soviet physicists
Jacob, Lise; François Jacob's wife
Jacob,* François (1920–2013). French biomedical scientist; Nobel laureate (1965)
Jerne, Niels K. (1911–1994). Danish immunologist; Nobel laureate (1984)
John, Fritz (1910–1994). German-American mathematician
Johnson, Lyndon B. (1908–1973). 36th president of the United States (1963–1969)
Kadanoff, Leo (1937–). American physicist
Kalckar, Fritz (1910–1938). Danish physicist; Herman Kalckar's brother
Kalckar, Herman (1908–1991). Danish-American biochemist; Fritz Kalckar's brother
Kandel, Eric R. (1929–). Austrian-born American biomedical scientist; Nobel laureate (2000)
Kapitsa, Piotr L. (1894–1984). Soviet physicist; Nobel laureate (1978)
Karle,* Isabella L. (1921–). American chemist; Jerome Karle's wife
Karle,* Jerome (1918–2013). American chemist; Nobel laureate (1985). Isabella Karle's husband
Katzir, Ephraim (1916–2009). Israeli politician and biophysicist
Keats, John (1795–1821). English poet
Kendall, Henry W. (1926–1999). American physicist; Nobel laureate (1990)
Kendrew, John C. (1917–1997). British biochemist; Nobel laureate (1962)
Kennedy, John F. (1917–1963). 25th US president (1961–1963)
Kepler, Johannes (1571–1630). German mathematician, astronomer, author
Ketterle,* Wolfgang (1957–). German-American physicist; Nobel laureate (2001)

Khariton, Yulii (1904–1996). Soviet physicist; head of the first Soviet nuclear weapons laboratory, Arzamas-16
Khorana, Har Gobind (1922–2011). Indian (Pakistani)-Canadian molecular biologist; Nobel laureate (1968)
Kilby, Jack S. (1923–2005). American engineer; Nobel laureate (2000)
King, Martin Luther (1929–1968). American clergyman, civil rights movement leader
Kirkwood, John G. (1907–1959). American chemist and physicist
Kissinger, Henry A. (1923–). American politician; President Nixon's secretary of state; Nobel laureate (1973)
Kleppner, Daniel (1932–). American physicist
Klug,* Aaron (1926–). British biophysicist and molecular biologist; Nobel laureate (1982)
Koestler, Arthur (1905–1983). Hungarian-British journalist and author
Kohler, Elmer Peter (1865–1938). American chemist
Köhler, Georges J. F. (1946–1995). German biologist; Nobel laureate (1984)
Kohn, Walter (1923–). Austrian-born American physicist; Nobel laureate (1998)
Konopka, Ronald (Ron) J. Seymour Benzer's former PhD student
Kornberg, Ken; American architect; Arthur Kornberg's son
Kornberg, Roger D. (1947–). American biochemist; Nobel laureate (2006). Arthur Kornberg's son
Kornberg, Sylvy, née Levy; American biochemist; Arthur Kornberg's first wife
Kornberg, Thomas B. (1948–). American biochemist; Arthur Kornberg's son
Kornberg,* Arthur (1918–2007). American biomedical scientist; Nobel laureate
Koussevitzky, Moshe (1899–1966) Lithuanian-born American Jewish cantor and vocalist
Kroemer, Herbert (1928–). German-American engineer and applied physicist; Nobel laureate (2000)
Kroto, Harold W. (1939–). British chemist; Nobel laureate (1996)
Kuhn, Thomas S. (1922–1996). American philosopher of science
Kurchatov, Igor V. (1903–1960). Soviet physicist; supreme leader of the Soviet nuclear program
Kuria, Ivan; Vladimir Prelog's teacher in Osijek, Croatia
Kuroda,* Reiko (1947–). Japanese chemist
Kurti,* Nicholas (1908–1998). Hungarian-British physicist
Landau, Lev D. (1908–1968). Soviet physicist; Nobel laureate (1962)
Langevin, Paul (1872–1946). French physicist
Laue, Max von (1879–1960). German physicist; Nobel laureate (1914)
Lauterbur,* Paul C. (1933–2007). American chemist; Nobel laureate (2003)
Lederberg,* Joshua (1925–2008). American biologist; Nobel laureate (1958)
Lee, Ho Wang; Korean virologist
Lee, Pyong Wuo; Korean virologist
Lee, Tsung-Dao (1926–). Chinese-American physicist; Nobel laureate (1957)
Lee,* Yuan Tseh (1936–). Taiwanese-American chemist; Nobel laureate (1986)
Leggett, Anthony (Tony) J. (1938–). British–American physicist; Nobel laureate (2003)
Lehn,* Jean-Marie (1939–). French chemist; Nobel laureate (1987)
Lenin, Vladimir I. (1870–1924). Russian communist revolutionary and dictator; first leader of the Soviet Union
Lennard-Jones, John E. (1894–1954). British physicist
Levene, Phoebus A. (1869–1940). Lithuanian-born American biochemist

Levi, Giuseppe (1872–1965). Italian biomedical scientist; mentor of Salvador Luria, Renato Dulbecco, and Rita Levi–Montalcini
Levi, Primo (1919–1987). Italian chemist and author
Levi-Montalcini,* Rita (1909–2012). Italian biomedical scientist; Nobel laureate (1986)
Lewis,* Edward B. (1918–2004). American biologist; Nobel laureate (1995)
Lewis, Gilbert N. (1875–1946); American physical chemist
Lindemann, Frederick A., Lord Cherwell (1886–1957). British physicist
Lipscomb,* William N. (1919–2011). American chemist; Nobel laureate (1976)
Long, Franklin A. (1910–1999). American chemist and arms control advisor
Lukes, Rudolf. Vladimir Prelog's mentor in Prague
Luria, Salvador (1912–1991). Italian-American biologist; Nobel laureate (1969)
Lwoff, André (1902–1994). French biologist; Nobel laureate (1965)
Lysenko, Trofim D. (1898–1970). Soviet charlatan agronomist
Ma, Yo-yo (1955–). American cellist
MacLeod, Colin M. (1909–1972). American biomedical scientist
Mandelbrot,* Benoit B. (1924–2010). Polish-French-American mathematician
Mansfield,* Peter (1933–). British physicist; Nobel laureate (2003)
Marshall, John (1917–1997). American physicist
Martin, A. J. P. (1910–2002). British chemist; Nobel laureate (1952)
Mason,* Stephen (1923–2007). English science historian
Maxwell, James Clerk (1831–1879). British physicist
McCarty,* Maclyn (1911–2005). American biomedical scientist
McMillan, Edwin M. (1907–1991). American physicist; Nobel laureate (1951)
Meitner, Lise (1878–1968). Austrian-German physicist
Mendeleev, Dmitrii I. (1834–1907). Russian chemist
Merrifield,* Bruce (1921–2006). American chemist; Nobel laureate (1984)
Meselson,* Matthew (1930–). American biologist
Michel, Hartmut (1948–). German biochemist; Nobel laureate (1988)
Milstein,* César (1927–2002). Argentinean-born British biomedical scientist; Nobel laureate (1984)
Mirsky, Alfred E. (1900–1974). American molecular biologist
Molina, Mario J. (1943–). Mexican-American chemist; Nobel laureate (1995)
Moncada,* Salvador (1944–). Honduran-born British biomedical scientist
Monod, Jacques (1910–1976). French biologist; Nobel laureate (1965)
Mössbauer,* Rudolf (1929–2011). German physicist; Nobel laureate (1961)
Mott, Nevill F. (1905–1996). British physicist; Nobel laureate (1977)
Müller-Hill,* Benno (1933–). German geneticist and author
Murad III (1546–1596). Sultan of the Ottoman Empire (1576–1612)
Murray, Joseph (1919–2012). American surgeon; Nobel laureate (1990)
Mussolini, Benito (1883–1945). Italian fascist dictator
Nathans, Daniel (1928–1999). American biologist; Nobel laureate (1978)
Ne'eman,* Yuval (1925–2006). Israeli physicist and military and political leader
Nernst, Walther H. (1864–1941). German chemist and physicist; Nobel laureate (1920)
Nesmeyanov, Aleksandr N. (1899–1980). Soviet chemist and science administrator
Neuberger, Albert (1908–1996). German-British biomedical scientist
Neumann, John von (1903–1957). Hungarian-American mathematician
Newton, Isaac (1642–1727). British scientist
Nirenberg, Marshall W. (1927–2010). American biochemist; Nobel laureate (1968)

Nixon, Richard M. (1913–1994). 37th US president (1969–1974)
Norrish, R. G. W. (1897–1978). British chemist; Nobel laureate (1967)
Nunn, Samuel A. (1938–). American politician; US senator
Nurse,* Paul (1949–). British biologist and science administrator; Nobel laureate (2001)
Nüsslein-Volhard,* Christiane (1942–). German biologist; Nobel laureate (1995)
O'Keefe, Georgia (1887–1986). American painter; Alfred Stieglitz's wife
Ochoa, Severo (1905–1993). Spanish-American biochemist; Nobel laureate (1959)
Olah,* George A. (1927–). Hungarian-American chemist; Nobel laureate (1994)
Oliphant,* Mark (1901–2000). Australian physicist
Oppenheimer, J. Robert (1904–1967). American physicist, first director of the Los Alamos National Laboratory
Osawa, Eiji (1935–). Japanese chemist
Osheroff, Douglas D. (1945–). American physicist; Nobel laureate (1996)
Panofsky,* Wolfgang K. H. (1919–2007). American physicist
Pasteur, Louis (1822–1895). French microbiologist
Pauli, Wolfgang (1900–1958). Austrian-Swiss physicist; Nobel laureate (1945)
Pauling,* Linus (1901–1994). American chemist; twice Nobel laureate (1954; 1963 for 1962)
Pavlov, Ivan P. (1849–1936). Russian physiologist; Nobel laureate (1904)
Pedersen, Charles J. (1904–1989). American chemist; Nobel laureate (1987)
Penrose,* Roger (1931–). British physicist and author
Penzias,* Arno A. (1933–). American astrophysicist; Nobel laureate (1978)
Perutz,* Max F. (1914–2002). Austrian-British biochemist; Nobel laureate (1962)
Phillips, William D. (1948–). American physicist; Nobel laureate (1997)
Pickering, William H. (1910–2004). American physicist and engineer
Planck, Max (1858–1947). German physicist; Nobel laureate (1919 for 1918)
Polanyi, Michael (1891–1976). Hungarian-British chemist and philosopher; John C. Polanyi's father
Polanyi,* John C. (1929–). British-Canadian chemist; Nobel laureate (1986)
Politzer, H. David (1949–). American physicist; Nobel laureate (2004)
Polkinghorne,* John C. (1930–). British physicist and Anglican priest
Pollister, A. W. American biochemist
Pople,* John A. (1925–2004). British-American chemist; Nobel laureate (1998)
Porter,* George (1920–2002). British chemist; Nobel laureate (1967)
Prelog,* Vladimir (1906–1998). Croatian-Swiss chemist; Nobel laureate (1975)
Priestley, Joseph (1733–1804). English natural philosopher
Primakoff, Henry (1914–1983). American physicist; Mildred Cohn's husband
Pritchard,* David E. (1941–) American physicist
Prokhorov, Aleksandr M. (1916–2002). Soviet physicist; Nobel laureate (1964)
Prusiner, Stanley B. (1942–). American biomedical scientist; Nobel laureate (1997)
Purcell, Edward M. (1912–1997). American physicist; Nobel laureate (1952)
Rabi, Isidor I. (1898–1988). American physicist; Nobel laureate (1945 for 1944)
Ramakrishnan, Venkatraman (1952–). Indian–British biologist; Nobel laureate (2009)
Ramsey,* Norman F. (1915–2011). American physicist; Nobel laureate
Rayleigh, John W., Lord Strutt (1842–1919). British physicist; Nobel laureate (1904)
Rich, Alexander (1924–). American biophysicist
Robbins,* Frederick C. (1916–2003). American pediatrician and biomedical researcher; medical administrator; Nobel laureate (1954)

Roberts, Louis (1913–1995) ; American physicist
Roberts, Richard J. (1943–) ; American biologist; Nobel laureate (1993)
Robinson, Robert (1886–1975). British chemist; Nobel laureate (1947)
Rockefeller Mauzé, Abby (1903–1976). American philantropist
Rose, Irwin A. (1926–). American biologist; Nobel laureate (2004)
Rose, Leonard (1918–1984). American cellist and pedagogue
Rowland,* F. Sherwood (1927–2012). American chemist; Nobel laureate (1995)
Rubens, Peter Paul (1577–1640). Flemish painter
Rubin,* Vera C. (1928–). American astrophysicist
Rudolf II (1552–1612). Holy Roman Emperor (1576–1612)
Russell, Bertrand (1872–1970). British philosopher, Nobel laureate (1950)
Rutherford, Ernest (1872–1937). British physicist; Nobel laureate (1908)
Ruzicka, Leopold (1887–1976). Croatian-Swiss chemist; Nobel laureate (1939)
Ryan, Francis. Joshua Lederberg's mentor at Columbia College
Ryle, Martin (1918–1984). British astrophysicist; Nobel laureate (1974)
Sabin, Albert B. (1906–1993). American microbiologist
Salam, Abdus (1926–1996). Pakistani-British physicist; Nobel laureate (1979)
Salk, Jonas E. (1914–1995). American virologist
Sanger,* Frederick (1918–). British biochemist; twice Nobel laureate (1958 and 1980)
Schmidt, Helmut (1918–). German politician; West German chancellor (1974–1982)
Schmidt-Ott, Friedrich (1860–1956). German science administrator
Schrödinger, Erwin (1887–1961). Austrian physicist; Nobel laureate (1933)
Seaborg,* Glenn T. (1912–1999). American chemist; Nobel laureate (1951)
Segré, Emilio G. (1905–1989). Italian-American physicist; Nobel laureate (1959)
Semenov,* Nikolai N. (1896–1986). Soviet chemical physicist; Nobel laureate (1956)
Shalnikov, Aleksandr I. (1905–1986). Soviet physicist
Shechtman,* Dan (1941–). Israeli materials scientist; Nobel laureate (2011)
Shoenberg, David (1911–2004). British physicist
Sigler, Paul B. (1934–2000). American molecular biologist
Simon, Franz (Francis) (1893–1956). German-British physicist
Skou,* Jens Chr. (1918–). Danish biomedical scientist; Nobel laureate (1997)
Smalley, Richard E. (1943–2005). American chemist and physicist; Nobel laureate (1996)
Smith, Hamilton O. (1931–). American biomedical scientist; Nobel laureate (1978)
Smithies, Oliver (1925–). British-American biologist; Nobel laureate (2007)
Stahl, Franklin (1929–). American biologist
Stalin, Iosif V. (1978–1953). Soviet dictator
Steitz, Thomas A. (1940–). American biophysicist; Nobel laureate (2009)
Stent,* Gunther S. (1924–2008). American molecular biologist
Stieglitz, Alfred (1864–1946). American photographer; Georgia O'Keeffe's husband
Stieglitz, Julius (1867–1937). American chemist
Strassmann, Fritz (1902–1980). German chemist; co-discoverer of nuclear fission
Straus, Eugene (1940–2011). American biomedical scientist and author
Sturkey, Lorenzo. American electron diffraction specialist
Sulston,* John E. (1942–). British biochemist; Nobel laureate (2002)
Synge, R. L. M. (1914–1994). British chemist; Nobel laureate (1952)
Szilard, Leo (1898–1964). Hungarian-American scientist and political activist
Sztehlo, Gabor (1909–1974). Hungarian Lutheran minister

Tamm, Igor E. (1895–1971). Soviet physicist; Nobel laureate (1958)
Tatum, Edward L. (1909–1975). American biochemist; Nobel laureate (1958)
Taube, Henry (1915–2005). Canadian-American chemist; Nobel laureate (1983)
Taylor, Joseph H. (1941–). American astrophysicist; Nobel laureate (1993)
Taylor, Richard E. (1929–). American physicist; Nobel laureate (1990)
Telegdi,* Valentine L. (1922–2006). Hungarian-American physicist
Teller,* Edward (1908–2003). Hungarian-American physicist
Temin, Howard M. (1934–1994). American virologist; Nobel laureate (1975)
Tisza, Laszlo (1907–2009). Hungarian-American physicist
Tonegawa, Susumu (1939–). Japanese molecular biologist; Nobel laureate (1987)
Townes,* Charles H. (1915–). American physicist; Nobel laureate (1964)
Tuppy, Hans (1924–). Austrian biochemist and administrator
Udenfriend, Sidney (1918–2001). American biochemist
Ulam, Stanislaw (1909–1984). Polish-American mathematician
Urey, Harold C. (1893–1981). American chemist; Nobel laureate (1934)
Van Vleck, John H. (1899–1980). American physicist; Nobel laureate (1977)
Varmus,* Harold E. (1939–). American biomedical scientist and administrator; Nobel laureate (1989)
Varshavsky,* Alex (1946–). Russian-American chemist and biomedical scientist
Veltman,* Martinus, J. G. (1931–). Dutch physicist; Nobel laureate (1999)
Venter, J. Craig (1946–) ; American biologist and entrepreneur
Vernon–Jones, V. S. Senior Tutor at Magdalene College, Cambridge, UK
von Weizsäcker, Carl Friedrich (1912–2007). German physicist and philosopher
Warburg, Otto (1883–1970). German biomedical scientist; Nobel laureate (1931)
Watson,* James D. (1928–). American biologist; Nobel laureate (1962)
Weber, Klaus (1936–). German biomedical scientist
Weigle, Jean-Jacques (1901–1968). Swiss biologist and physicist
Weil–Malherbe, H. German refugee biochemist in the UK and USA
Weinberg,* Steven (1933–). American physicist; Nobel laureate (1979)
Weinrich, Marcel; American physicist
Weissbach, Herbert. American biochemist
Weissmann, Julie. Charles Weissmann's wife
Weissmann,* Charles (1931–). Swiss-American biomedical scientist
Wellcome, Henry (1853–1936). British pharmacist and philanthropist
Weller, Thomas H. (1915–2005). American virologist; Nobel laureate (1954)
Westheimer, Jeanne; Frank Westheimer's wife
Westheimer,* Frank H. (1912–2007). American chemist
Wheeler, Janet. John A. Wheeler's wife
Wheeler,* John A. (1911–2008). American physicist
Whitehead, A. N. (1861–1947). English mathematician and philosopher
Wieman, Carl E. (1951–). American physicist; Nobel laureate (2001)
Wieschaus, Eric F. (1947–). American molecular biologist; Nobel laureate (1995)
Wiesner, Jerome (1915–1994). American educator and science administrator
Wigner, Eugene P. (1902–1995). Hungarian-American physicist; Nobel laureate (1963)
Wilczek,* Frank (1951–). American physicist; Nobel laureate (2004)
Wilde, Oscar (1854–1900). Irish writer and poet
Wilhelm IV, Landgrave of Hesse (1532–1592). patron of arts and sciences

Wilkins, Maurice H. F. (1916–2004). British biophysicist; Nobel laureate (1962)
Wilson, E. Bright (1908–1992). American physical chemist
Wilson, Edward O. (1929–). American biologist and author
Wilson,* Kenneth G. (1936–). American physicist; Nobel laureate (1982)
Wilson, Robert W. (1936–); American physicist; Nobel laureate (1978)
Wittig, Georg (1897–1987). German chemist; Nobel laureate (1979)
Wittmann, H. G.; German molecular biologist
Wolfrum, Jürgen (1939–). German physical chemist
Wollman, Élie (1917–2008). French biologist
Woodward, Robert B. (1917–1979). American chemist; Nobel laureate (1965)
Wooley, Wayne; American chemist
Wu, Chien-Shiung (1912–1997). Chinese-American physicist
Yalow, Aaron (1920–1992). American physicist; Rosalyn Yalow's husband
Yalow,* Rosalyn S. (1921–2011). American physicist; Nobel laureate (1977)
Yang, Chen Ning (1922–). Chinese-American physicist; Nobel laureate (1957)
Yeats, W. B. (1865–1939). Irish poet; Nobel laureate (1923)
Yonath,* Ada (1939–). Israeli biologist; Nobel laureate (2009)
Zare,* Richard (1939–). American chemist
Zeldovich, Yakov B. (1914–1987). Soviet physicist
Zewail,* Ahmed (1946–). Egyptian-American chemist; Nobel laureate (1999)

INDEX

Asterisks indicate names of interviewees excerpts from whose conversations are presented in this volume.

Abelson, Philip H., 162
Abraham Lincoln High School, 173
Abrikosov, Alexei, A., 35
Academic Assistance Council, 43
accelerators, 40, 72, 106, 107
acoustic noise, 303, 304
administrative positions, 243, 340, 341
Aerosol Age, 211
aesthetic value, 117, 184
Age of Discontinuity, The (Drucker), 122
Aguet, Michel, 351
Akhmatova, Anna, 345
Aleksandrov, Anatoly P., 4
Alferov,* Zhores I., 3–5
Allied Chemicals, 211
alpha-helix model, 326, 327
Alpher, Ralph, 79, 91, 92
Alzheimer's disease, 147, 267
American Chemical Society (ACS), 94, 161
Amherst College, 340
Aminoff Prize, 168
Anderson,* Philip W., 6–8
Anglican Church, 81
Ann, Princess, 206
Antarctica, 231, 232
antibiotics, 151, 218, 229, 248
antibodies, 218, 268, 310
anti-communism, 99, 100, 101
anti-Semitism, 56, 58, 59, 100, 174, 262, 296, 297
Apprentice to Genius (Kanigel), 257
Arber,* Werner, 239–241
Archimedes, 85
Aristotle, 115, 335
arms control, 72, 200, 308
Asilomar meetings, 240, 248

Astbury, William T., 326, 327
Astrobiological Institute (of NASA), 251
astronomy, 11, 48, 79, 91, 104, 188
astrophysics, 8, 48, 49, 53, 92, 108, 110
asymmetry. *See* symmetry
atomic bombs, 7, 32, 34, 44, 60, 114, 176, 177, 200, 222, 223
atomic clocks, 87
Auschwitz, 272–274
Australian National University, 69
autoimmune reactions, 268
Avery Oswald T., 153, 293, 294, 305, 306, 351
Ax, Emanuel, 287
Axelrod, Julius, 258

Bacon, Francis, 188
bacteria, bacteriology, 145, 228, 229, 232, 246, 248, 280, 283, 292, 352, 306, 329
Baltimore,* David, 242–244
Banting, Frederick G., 269
Bárány, Anders, 12
bases in DNA, 129, 130, 218, 274, 294
basic research, 63, 220, 259, 332, 333
Basov, Nikolai G., 102
Bastiansen, Otto, 196
Baylen, Sarah, 127
Beadle, George W., 294
beauty, 83, 209, 297, 320
Bekesy, George, 195
Bell Burnell,* Jocelyn, 9–12, 49
Bell Laboratories, 6, 78, 102
Ben Gurion, David, 262
Benzer,* Seymour, 245–246
Berg,* Paul, 247–249
Berlin colloquia (Physik Kolloquia), 42, 57

Bernard, Claude, 267
Berson, Solomon, 353–355
Bethe, Hans, 32, 120
Big Bang model, 49, 79, 104, 108
big science, 39
bioconversion, 219
biography, autobiography, 43, 100, 296, 334, 344, 345
biohazards, 248
biological evolution, 240, 241
biological weapons, 307, 308
biomarkers, 232
biosphere, 142
biostatistics, 299, 329
Birkbeck College, 282
birth control, 147
Bishop, J. Michael, 340
Blaschko, Hermann, 276
blood poisoning, 136
Blumberg,* Baruch S., 250–252, 267
Bohr, Niels, 112–114, 121, 222, 335
Bose-Einstein Condensation (BEC), 53–55
Boston Children's Hospital, 328
bovine spongiform encephalopathy ("mad cow disease"), 267
Bragg, W. Lawrence, 206, 325, 327
Brahe, Tycho, 188
brain
   function, 7, 249, 267, 300, 301, 347, 351, 352
   at the frontier of research, 25, 76, 77, 261
   and memory, 218, 249
   and Parkinson's disease, 256, 257, 259, 275–278
   modelling, 259
   and consciousness, 264
Bréchignac,* Catherine, 13–15
Breit, Gregory, 43, 44, 112, 113
Brenner,* Sydney, 253–255, 265, 338
Brockway, Lawrence, 169
Brodie, Bernard B., 256–258
Bronx Veterans Administration Medical Center, 353, 354
Brookhaven National Laboratory, 41, 354
Brooklyn College, 245
Brown,* Herbert C., 125–128, 195
Brownlee, George, 216

bubble chamber, 38, 40
Buckingham Palace, 206
buckminsterfullerene, 158, 197
Budapest Technical University, 100, 193
Büeler, Hansruedi, 351
Burnet, Frank Macfarlane, 268, 330
Burroughs Wellcome, 149, 150
Burroughs, Silas M., 150

Cabot Lodge, Henry, Jr., 309
Cairns, John F., 287
California Institute of Technology. *See* Caltech
Calne, Roy Y., 150
Caltech (California Institute of Technology)
   Baltimore, 242
   Benzer, 245, 246
   Gajdusek, 266
   Garwin, 30
   Glaser, 38
   Koch, 265
   Lewis, 299, 300
   Lipscomb, 184, 186
   Mandelbrot, 59, 60
   Meselson, 307
   Mössbauer, 62
   Panofsky, 72
   Politzer, 116
   Pritchard, 84
   Stent, 334
   Telegdi, 96
   Townes, 102
   Varshavsky, 343
   Wilson (K. G.), 119
   Zewail, 234
Cambridge University (Cambridge, UK)
   Anderson, 6
   Bell Burnell, 9, 10
   Brenner, 253
   Chibnall, 214
   Crick, 263
   Dirac, 203
   Dyson, 23
   Gilbert, 270
   Goldhaber, 41, 43
   Hewish, 9, 48
   Klug, 282

Mason, 187
Milstein, 310
Norrish, 205
Oliphant, 70
Penrose, 75
Polkinghorne, 81, 82
Pople, 202, 203, 204
Ramsey, 87
Rutherford, 70
Sanger, 213, 214
Sulston, 337
Watson, 346
cancer, 27, 147, 189, 248, 251, 261, 339, 341, 342
Cantico di una vita (Levi-Montalcini), 296
Cantor's Dilemma (Djerassi), 147, 148
Capecchi, Mario, 351
carbocation chemistry, 193–195
Carlsson,* Arvid, 256–259, 276
Carnegie Institution, 90
Carnegie Mellon University, 202, 203
Carter, Jimmy, 73
Case Institute of Technology, 38, 290
Case Western University, 328, 329
catalysis, 133, 140, 225, 226
Catholic Church, 68, 79, 188
Cavendish Laboratory, 41, 43, 48, 263, 327
Celera company, 338
cell biology, 285, 295, 307, 319, 343
censorship, 36
Centre National de la Recherche Scientifique. *See* CNRS
CERN, 96, 106, 107
Chadwick, James, 43
chain reactions, 221, 222
Chain, Boris, 136
Chandrasekhar, Subrahmanyan, 43
Chargaff,* Erwin, 129–131, 293, 294
Charlemagne, 188
Charles, Prince of Wales, 347, 348
Chemical Evolution (Mason), 187
chemical reactions, 155, 165, 166, 205, 234
chemical structures, 196, 198
Chemistry of Penicillin, The, 136
chemotherapy, 218
Cheronis, Nicholas D., 127
Chiang, Kai-shek, 179, 180

Chibnall, Albert C., 214, 215
Chirkov Nikolai M., 222
chlorofluorocarbons (CFCs), 142, 211, 212
chromatography, 214–216
chromosomes, 254, 280, 300, 301
Churches, 36, 68, 79, 81, 188
Churchill, Winston, 35
Ciba Pharmaceutical, 146, 208
Ciechanover,* Aaron, 260–262, 273, 344
Citation Index, 156, 172
City College of New York, 78, 173, 286, 354
civilization, 156, 177, 267
Clarendon Laboratory, 56
classics, 271
claustrophobic effect, 303
Clemett, Simon, 231
Clinton, William (Bill) J., 73
cloning, 218, 271, 310, 351
cloud chamber, 39
CNRS (Centre National de la Recherche Scientifique), 13, 15
Cohen, Stanley, 295
Cohn,* Mildred, 132–134
Cold Spring Harbor Laboratory (CSHL), 346, 347
Colegio Nacional in Bahía Blanca, 310
Collège de France, 181
Cologne University, 316
Columbia University
  Blumberg, 250
  Chargaff, 129
  Cohn, 132
  Gajdusek, 268
  Glaser, 39
  Hammett, 225
  Heidelberg, 268
  Hoffmann, 165, 166
  Kornberg (T.), 287
  Lederberg, 292–294
  Penzias, 78
  Pollister, 293
  Rabi, 134
  Ramsey, 87
  Ryan, 294
  Townes, 102
  Urey, 134

Columbia University (*Cont.*)
  Varmus, 340
  Weinberg, 109
  Westheimer, 225
  Zare, 230
Columbus, Christopher, 189
combinatorial chemistry, 152, 154, 342
Common Thread, The (Sulston and
    Ferry), 338
communism, 100, 180
complementarity, 246
complexity, 8, 182, 229, 241, 261
computational science, 24, 67, 38, 156, 159,
    198, 202, 203, 224, 342
Conant, James B., 224–226
consciousness, 75, 76, 83, 264, 336
Consilience (Wilson), 8
contraceptives, 146, 147
controversies, 28, 76, 99, 194, 195, 212,
    240, 284
Conway,* John H., 16–18
Cooper Union College, 355
cooperativity, 139
Copernicus, Nicolaus, 111, 121
Cormack, Alan M., 284
Cornell Medical School, 132
Cornell University, 19, 20, 23, 90, 109, 119,
    120, 165, 308
Cornell, Eric A., 53, 55
Cornforth,* John W., 135–137
cosmic rays, 38, 79
cosmology, 8, 79, 92, 113
Cotton, F. Albert, 228
Cotzias, George, 277
Coulombic forces and interactions, 14
Courant Institute, 185
Cram,* Donald J., 138–140, 181
Crane Junior College, 126, 127
creativity, 121, 220, 258, 268, 269, 320
Creutzfeldt-Jacob disease, 267
Crick, Odile, 176, 264
Crick,* Francis, 263–265
  greatness of, 121, 122, 130, 218
  and Watson, 121, 122, 293, 305, 346
  criticism of, 130
  in Tokyo, 176
  discovering the double helix, 218, 293,
    305, 346

  unsuccessful in crystallizing
    ribosomes, 229
  on the right to terminate life, 263
  on medical care for old people, 264, 265
  pupils and coworkers of, 265
  autobiography of, 344
Cronin, Alexander, 86
Crutzen,* Paul J., 141–142, 211
crystallography, 144, 171, 172, 176, 197
  classical, 93, 197
  crystal structures, 139, 172, 196, 198, 207
  crystallographic electron microscopy, 283
  protein crystallography, 227, 228, 325
  quasicrystals, 93–95, 197
  ribosome structure, 229
culture
  American, 236
  for peaceful coexistence, 183
  in Jewish life, 261, 262, 297, 312
  in stone-age, 266, 267
  Japanese, 177
  Middle Eastern, 236
  of scientists, 148
  preservation of, 348
  science as part of , 111, 148
  two cultures, 146, 148
curiosity, 92, 315, 344, 345
Curl, Robert F., 197
cyclotrons, 27, 28, 39
cystic fibrosis, 348

Dalton, John, 188
Darkness at Noon (Koestler), 101
Darwin, Charles, 121, 122, 154, 188
Davy, Humphry, 206
Dawson, Joan M., 289
deafness, 135–137, 218
debates, 36, 73, 199, 200, 240
Degkwitz, R., 276
Deguchi, Takeo, 314
Deisenhofer,* Johann, 143–145
Delbrück, Max, 245, 246, 266, 334
developmental biology, 287, 322
diabetes, diabetics, 191, 259, 269, 277, 278
Dirac, Paul A. M., 43, 121, 203
disarmament, 72, 308
discoveries, 80, 83, 103, 147, 149, 280, 300,
    314, 320, 344, 345

astrophysical, 49
asymptotic freedom in strong
    interactions, 116
base equivalence in DNA, 130
biological synthesis of nucleic acids, 286
breast cancer genes, 339
buckminsterfullerene, 160, 197
cellular origin of retroviral
    oncogenes, 340
chemical reactions, 194, 222
contraceptive pill, 147
cosmic microwave background radiation,
    79, 80, 92
covalent bond, 186
dark matter, 90
depletion of the ozone layer, 211
DNA as genetic material, 305, 306
double helix, 263, 346, 347
drugs, 149
elements and isotopes, 217
fiber structures, 326
genetic control of embryonic
    development, 300, 322
genetic regulation of organ
    development, 338
helium-3, 70
hepatitis B virus, 250
insulin, 259, 269, 277, 278
ion-transporting enzymes, 332
L-DOPA treatment, 258, 259, 277, 278
magnetic resonance imaging, 302
mathematical, 16
monoclonal antibodies, 310
Mössbauer effect, 62
nerve growth factor, 295, 296
New World, 189
NO, 314
partition chromatography, 214
primordial plutonium, 162
protein structures, 327
pulsars, 9, 12, 48, 49
quasicrystals, 93, 197
radioimmunoassay, 354, 355
regulators of the cell cycle, 319
restriction enzymes, 239
serotonin depletion, 257
structure of photosynthesis reaction
    center, 144

supersymmetry, 110
tritium, 70
ubiquitin-mediated protein degradation,
    260, 273, 343
X-rays, 49
diseases, 147, 218, 259, 267, 300, 348
    Alzheimer's, 147, 267
    bovine spongiform encelaphopathy,
       267
    brain diseases, 277, 278, 352
    cancer, 147
    cardiovascular, 147
    Creutzfeldt-Jakob, 267
    Down's, 300
    hyperthyroidism, 306
    infectious, 218, 250, 267, 329, 350
    inflammation, 147
    kuru, 266, 352
    nervous system, 266
    Parkinson's, 257–259, 275–278
    prion, 351
    scrapie, 351
    smallpox, 348
    Viliuisk encephalomyelitis, 25
Djerassi,* Carl, 146–148
DNA, 153, 215, 247, 254, 286
    as natural product, 153, 293
    base equivalence, 129, 294
    chemistry, 153, 294
    function, 121, 218, 280, 293, 294, 301,
       305, 306, 351
    organism-specificity, 129, 130, 254
    polymerase, 287
    recombinant technology, 218, 240, 243,
       247, 347
    repair, 287
    replication, 307
    sequencing, 145, 215, 271, 342
    structure, 92, 293, 338, 346, 347
DNA Doctor, The (Hargittai), 347
Doctorow, E. L., 344
DOPA, L-DOPA, dopamine, 256–259,
    275–277
double helix, 121, 130, 263, 338, 346
Double Helix, The (Watson), 346, 347
Dresselhaus,* Mildred S., 19–22
Drucker, Peter, 122
drug companies, 149, 338

drug research, 149, 150, 218, 257, 258, 338, 342
drugs, 147, 149–151, 277
Duke University, 102, 130
Dulbecco, Renato, 242, 246, 297
DuPont, 211
Dyson,* Freeman J., 23–25, 308

Earth, 11, 17, 51, 52, 63, 73, 79, 156, 157, 163, 231, 232, 241, 252
École Normale, 14
École Polytechnique, 59
EDRF (endothelium-derived relaxing factor), 313, 314
Eigen, Manfred, 205
Einstein, Albert, 26, 42, 98, 104, 113, 121, 122, 344
Eisenhower, Dwight D., 73, 74
electromagnetism, 304
electron microscopy, 232, 283
elementary particles, 39, 41, 65, 66, 81, 82, 109, 221
Elion,* Gertrude B., 149–151
Elson, D., 229
embryonic development, 300, 322, 335, 351
Enders, John F., 328–330
endothelium-derived relaxing factor. *See* EDRF
environmental issues, 141, 142, 156, 212, 219, 220, 248
enzymes, 218, 226, 239, 279, 280, 294, 314, 332
Ernst, Richard, 179, 290
Eschenmoser,* Albert, 152–154
ETH (Swiss Federal Institute of Technology, Zurich), 96, 152, 207, 239
ethical considerations, 171, 219, 240, 300, 324
Evenson, Robert E., 122
evolution, 6, 8, 83, 139, 241
extraterrestrial intelligence, 51, 163

Fairlie, David B., 66
Faraday, Michael, 206, 255
Feinberg, Evgenii L., 36
femtosecond spectroscopy, 234
Fermi, Enrico, 27, 28, 42, 44, 98
Ferry, Georgina, 338

Feynman, Richard, 25, 107, 112, 113, 121
Fisher, Michael E., 120
fission phenomena, 14, 31, 112, 162
Fleming, Alexander, 136
Florey, Howard W., 136
Florida State University, 86
fossil fuel, 142
Fothergill, Leroy D., 308
Fowler, Ralph H., 43
Fox Chase Cancer Center, 250, 251, 260, 272, 273
fractals, 59–61
Fraenkel, George K., 166
Francis Crick Institute (UK), 319
Franklin, Rosalind, 283, 284, 347
Frederick II, 188
Freund, Peter, 117
Friedman,* Jerome I., 26–29, 97
Frost, Robert, 345
Fukui,* Kenichi, 155–157, 166
fullerenes, 158, 160, 197
fundamental particles. *See* elementary particles
funding, 58, 127, 130, 191, 332, 333, 341
Furman University, 102

Gajdusek,* D. Carleton, 25, 250, 266–269, 352
galaxies, 48, 49, 52, 90, 91, 163
Galich, Aleksandr, 35
Galilei, Galileo, 79, 111, 188
GALLEX (Gallium Experiment), 63
Galpern,* Elena, 158–160, 197
games, 17, 79, 91, 136, 153, 268
Gamow, George, 79, 90–92, 113
Garwin,* Richard L., 28, 30–33
gauge theory, 117
Gell-Mann, Murray, 65, 66, 98, 107, 119, 121
gender discrimination, 323, 348, 349
gene expression, 271, 280, 341, 342, 351
gene therapy, 218
General Electric, 303
genes and behavior, 245, 249, 300
genetic code, 218, 229, 264
genetic defects, 218, 300, 301
genetic engineering, 24, 239–241, 300
genetic manipulation, 248, 280, 292
genetically modified food, 239, 240, 347

genetics and intelligence, 68, 247, 249, 300, 301, 318, 347
Geneva, 40, 96, 239
Genius in the Shadows (Lanouette), 43
genome, 218, 240, 241, 271, 280, 338, 339, 342
George Washington University, 91, 113
Georgetown University, 90, 91
Gilbert,* Walter, 270–271, 317, 318
Gilcrease, F. Wellington, 173
Ginzburg,* Vitaly L., 34–37
Glaser,* Donald A., 38–40
Glashow, Sheldon L., 109
Glaxo Wellcome, 149
God, 7, 36, 69, 71, 83, 104, 105, 153, 240, 261
Gödel's theorem, 17
Goeppert Mayer, Maria, 27
Goering, Hermann, 189
Goldhaber,* Maurice, 41–44, 354
Grand Unifying Principle, 51
Great Books, 127
Great Depression, 100, 126
Greengard, Paul, 256
Greytak, Thomas, 54, 55
Gross,* David, 45–47, 116, 117
growth factors, 191, 295, 297
Gruss, Peter, 351
Guggenheim Foundation, 114

Haber, Fritz, 57
Hadlow, William J., 352
hadron collider, 110
Hahn, Otto, 42, 162
Halford, Ralph S., 166
Hamaroff, Stewart, 76
Hamburger, Viktor, 296
Hammett, Louis P., 133, 225
Hartwell, Leland H., 319
Harvard University
  Anderson, 6
  Conant, 225
  Cram, 138
  Dresselhaus, 19
  Enders, 328
  Gajdusek, 266
  Gilbert, 270, 317, 318
  Hoffmann, 165, 166
  Karle (J.), 171, 173

Kissinger, 308, 309
Lee (Y. T.), 178
Lipscomb, 184, 185, 228
Meselson, 307–309
Müller-Hill, 317, 318
Pritchard, 84
Ramsey, 87, 88
Robbins, 328, 329
Watson, 348
Weinberg, 109, 110
Weller, 329
Westheimer, 224–226
Wilson (K. G.), 119
Yonath, 228
Zare, 230
Hebrew University, 45, 260, 272
Heidelberger, Michael, 268
Heilbron, Ian, 209
Heisenberg, Werner, 43, 100, 121
helium, 63, 69, 70
hepatitis, 250, 251, 268, 329
heredity, 300, 335
Herman, Helen, 92
Herman, Robert, 79, 91, 92
Herschbach, Dudley R., 178, 199
Hershey-Chase experiment, 293
Hershko,* Avram, 260, 272–274, 344
Hertz, Gustav L., 57
heterostructures, 3
Hewish,* Antony, 9, 48–49
Higgs particle, 66
Hilbert, David, 344
Hiroshima, 32
history, 46, 83, 104, 165, 166, 176, 177, 201, 261, 262, 316
history of science, 46, 83, 104, 165, 187–189, 335
  about authority, 187–189
  and the humanities, 166
  as Nobel Prizes rewriting it, 314
  chemistry, 225, 232, 233
  genetics in Germany, 316
  Nazi science, 316, 318
  of agricultural institutions, 121
  physics, 344
  research in Parkinson's disease, 277
  University of Tokyo, 175, 177
History of Science, A (Mason), 187

Hitchings, George H., 149, 150
Hitler, Adolf, 34, 35, 57, 133, 186, 309, 347
Hochstrasser, Robin M., 234, 235
Hoffman,* Darleane C., 161–164
Hoffmann,* Roald, 155, 165–167
Hofstadter, Robert, 29
Holley, Robert W., 216
Holocaust, 262, 273, 274
Hooft,* Gerardus 't, 50–52, 106, 107
Hopkins, Nancy, 348
Hornykiewicz,* Oleh, 275–278
Horvitz, H. Robert, 253, 338
Hounsfield, Godfrey N., 284
Hoyle, Fred, 104
Hubble, Edwin P., 104
Huber, Robert, 143, 144
Huffman, Wallace E., 122
Hulse, Russel A., 12
Hunt, R. Timothy, 319
Hunter College, 19, 132, 149, 353, 354
Hutchins, Robert M., 127
hydrocarbons, 142, 159, 160, 194, 211, 231, 233
hydrogen bomb (H-bomb), 7, 31–33, 73, 99, 112, 114

IBM, 30, 59
immunochemistry, 218, 305
immunology, 268
Imperial College (London), 205, 206
In Praise of Imperfection (Levi-Montalcini), 295, 296
Indiana University, 346
Innovation and Entrepreneurship (Drucker), 122
Institut Laue-Langevin (Grenoble), 62
Institut Pasteur (Paris), 279, 280, 334
Institute for Advanced Study (Princeton), 23
Institute for Theoretical Physics (Utrecht), 50
Institute of Brain Research (Vienna), 275
Institute of Cancer Research (London), 177
Institute of Cell Biology (Rome), 295
Institute of Chemical Physics (Leningrad), 221
Institute of Fundamental Chemistry (Kyoto), 155

Institute of Molecular Biology (Moscow), 343
Institute of Physical Problems (Moscow), 5
insulin, 214, 215, 259, 269, 277, 278
intelligence, 110, 247, 249, 268, 300, 301, 318, 347
    extraterrestrial, 10, 51
interferon, 351
intermolecular interactions, 182, 195
international cooperation, 4, 5
International Space Station, 251
interstellar molecules, 79, 80
Introduction to Scientific Research, An (Wilson), 119, 122
intuition, 7, 109–111, 156, 308
Ioffe Institute (St. Petersburg), 3–5
Ioffe, Abram F., 3, 4, 222, 223
Iowa State University, 161
isomorphous replacement, 325
Italian National Council of Research (CNR), 295

Jacob,* François, 176, 279–281
Jerne, Niels K., 310
Jesuits, 188
Jews in science, 67, 68, 189, 354
John, Fritz, 185
Johns Hopkins Applied Physics Laboratory (JPL), 91
Johns Hopkins University, 112, 305
Johnson & Johnson (company), 303
Johnson, Lyndon B., 309
Joint Institute for Laboratory of Astrophysics (JILA), 53
Judaism, 261
Jupiter, 52

Kadanoff, Leo, 120
Kalckar, Fritz, 113
Kalckar, Herman, 335
Kandel, Eric R., 256
Kapitsa, Piotr L., 4, 5
Karle,* Isabella L., 168–170, 172
Karle,* Jerome, 168–174, 179
Karolinska Institute, 260
Keats, John, 344
Kellogg Foundation, 315

Kendall, Henry W., 26
Kendrew, John C., 325, 326
Kennedy, John F., 73
Kepler, Johannes, 60, 79, 111, 121, 188
Ketterle,* Wolfgang, 53–55, 84, 85
KGB, 211
Khariton, Yulii, 4, 221
Khorana, Har Gobind, 293
Kilby, Jack S., 4
King, Martin Luther, 25
King's College (Cambridge, UK), 81, 253
King's College (London), 176, 177, 187
Kirkwood, John G., 266
Kissinger, Henry A., 67, 308
Kleppner, Daniel, 54, 55, 86
Klug,* Aaron, 229, 265, 282–285, 326
Koestler, Arthur, 99, 101
Kohler, Elmer Peter, 225
Köhler, Georges J. F., 310
Kohn, Walter, 202
Kondratiev, Viktor, 221
Konopka, Ronald J., 246
Kornberg, Roger D., 287
Kornberg,* Arthur, 286–288, 293
Kroemer, Herbert, 3
Kroto, Harold W., 197
Kuhn, Thomas S., 257
Kunsthaus (Zurich), 208, 209
Kuomintang (KMT), 180
Kurchatov, Igor V., 4
Kuria, Ivan, 208
Kuroda,* Reiko, 175–177
Kurti,* Nicholas, 56–58
kuru, 25, 266, 352
Kyoto Imperial University, 155

lac repressor, 316–318
Lancet, 352
Landau, Lev D., 4, 34, 100, 101, 117
Langevin, Paul, 56
lasers, 54, 55, 102, 103
Lauder School (Budapest), 261
Laue, Max von, 42, 57
Lauterbur,* Paul C., 289–291, 302
Lawrence Berkeley National Laboratory, 161, 217
Lawrence Livermore National Laboratory, 161
Lebedev Institute (Moscow), 5, 35
Lederberg,* Joshua, 292–294
Lee, Ho Wang, 268
Lee, Pyong Wuo, 268
Lee, Tsung-Dao, 28, 88, 107
Lee,* Yuan Tseh, 178–180, 199
Leggett, Anthony J., 35
Lehn,* Jean-Marie, 138, 181–183
Lenin, Vladimir I., 4, 35, 222
Lennard-Jones, John E., 203
Levene, Phoebus A., 294
Levi, Giuseppe, 297
Levi, Primo, 297
Levi-Montalcini,* Rita, 295–298
Lewis,* Edward B., 299–301, 322
Lewis, Gilbert N., 186, 217
life on Mars, 231–233
Lindemann, Frederick A., 58
Lipscomb,* William N., 165, 184–186, 228
London equation, 36
Los Alamos National Laboratory, 31, 32, 97, 101, 114, 161, 169
Lowell Observatory, 91
LSD, 257
Lukes, Rudolf, 208
Luria, Salvador, 246, 297, 346
Lwoff, André, 279, 280
Lysenko, Trofim D., 189, 300

Ma, Yo-yo, 287
MacLeod, Colin M., 305, 306
macromolecules, 176, 218
magnetic resonance, 87, 88, 289, 290, 302
magnetic resonance imaging. *See* MRI
Manchester University, 199
Mandelbrot,* Benoit B., 59–61
Manhattan Project, 43, 44, 69, 99, 112, 114, 168, 169, 171, 176
Mansfield,* Peter, 289, 302–304
Marshall, John, 28
Martian history, 233
Martin, A. J. P., 214, 215
Marxism-Leninism, 300
masers, 87, 103
Mason,* Stephen, 176, 187–189

mass spectrometry, 14, 86, 231
Massachusetts Institute of Technology.
    See MIT
Max Planck Institutes, 62, 141, 143, 144, 228
Max Planck Society, 323
Max von Laue–Paul Langevin Institute
    (Grenoble), 62
Maxwell's laws/equations, 186, 304
McCarthyism, 189
McCarty,* Maclyn, 293, 305–306
McMillan, Edwin M., 162, 217
medical research, 166, 339
Meitner, Lise, 42
Mellon Institute, 228
membrain proteins, 144
Memorial Sloan-Kettering Cancer Center
    (MSKCC), 340, 341
Mendeleev, Dmitrii I., 51
Merrifield,* Bruce, 190–192
Meselson,* Matthew, 307–309
Michel, Hartmut, 143, 144
microwave background radiation, 79
Mike test, 31–33
Milstein,* César, 179, 310–312, 326
Mirsky, Alfred E., 293
MIT (Massachusetts Institute of Technology)
    Baltimore, 242, 243
    Cotton, 228
    Dresselhaus, 19–22
    Friedman, 26
    Glaser, 39
    Hopkins, 348
    Ketterle, 53–54
    Pritchard, 84
    Ramsey, 88
    Rich, 207
    Townes, 102
    Varshavsky, 343
    Weinberg, 109
    Wilczek, 116, 117
    Zare, 230
models
    alpha-helix, 326, 327
    Big Bang model, 79
    brain, 259
    carbon materials, 159, 160
    DNA double helix, 92, 338

Higgs particle, 66
independent-particle, 114
liquid droplet, 114
molecular structure, 144, 145, 326
nuclear fission, 14
nuclear structure, 27
quark, 26
quasicrystals, 197
radiation implosion, 31
replication, 140, 293
self-reproducing machines, 254
Shell Model, 27
Standard Model, 51, 82, 117
transfer ribonucleic acid, 207
twinned crystal, 95
virus, 282
molecular beams, 88
molecular biology, 24, 46, 92, 218, 245, 263,
    270, 271, 280, 335, 342
molecular genetics, 239, 267, 287, 294
molecular medicine, 218
molecular structures, 154, 182, 198
Molina, Mario J., 141, 211
Moncada,* Salvador, 313–315
monoclonal antibodies, 218, 310
Monod, Jacques, 279, 280
Montreal Protocol, 211
moons, 51, 52, 79
Moscow State University, 196, 343
Mössbauer effect (Mössbauer
    spectroscopy), 62
Mössbauer,* Rudolf, 62–64
Mott, Nevill F., 6
Mount Sinai School of Medicine, 355
MRC (Medical Research Council), 263, 282,
    283, 285, 310, 325, 326, 337, 351
MRI, 290, 302, 303
Müller-Hill,* Benno, 316–318
multiple sclerosis, 269
muon decay, 28
Murad III, 188
Murderous Science (Müller-Hill), 316
Murray, Joseph, 150
Mussolini, Benito, 297
mutations, 246, 249, 294
myeloma, 268
Myriad Genetics (company), 339

Nagasaki, 32, 114
Nanking, 7
nano-sizes, 14
NASA, 212, 231, 232, 251, 267
Nathans, Daniel, 239
National Academy of Sciences (US), 21, 92, 170, 224, 248
National Bureau of Standards (now, National Institute of Standards and Technology, NIST), 86, 203
National Cancer Institute, 341
National Gallery of London, 209
National Heart Institute, 256
National Institutes of Health (NIH), 266, 267, 286, 291, 338, 340
National Physical Laboratory (UK), 203
National Research Council (US), 225
National Taiwan University, 178
National Youth Act (NYA), 134
natural products, 146, 153, 154
Nature of the Chemical Bond, The (Pauling), 198
Naval Research Laboratory (NRL), 6, 168, 171
Nazis, Nazi Germany, 43, 44, 56, 72, 100, 130, 146, 165, 193, 316, 318, 334, 347
Ne'eman,* Yuval, 65–68
neptunium, 162
Nernst, Walther H., 42, 57
nervous system, 256, 257, 266, 275, 284, 297
Nesmeyanov Institute (INEOS, Moscow), 158, 159
Nesmeyanov, Aleksandr N., 159
Neuberger, Albert, 213
neurology, 24, 25, 259, 285
Neurospora, 294
neutrinos, 63, 64, 110
neutron stars, 11, 49
New Coming, The (Levi-Montalcini), 297
New York City College System, 19, 78, 173, 286, 354
New York University, 112, 149, 350
Newton, Isaac, 47, 111, 121, 122, 203, 206
Newton's laws, 17, 121
NIH. *See* National Institutes of Health
Nirenberg, Marshall W., 179
NIST. *See* National Bureau of Standards

nitric oxide (NO), 313, 314
Nixon, Richard M., 308
NMR (nuclear magnetic resonance), 88, 290, 303
Nobel Prize
 as changing the awardee's life, 118, 134, 215, 328, 330, 341
 as influencving research directions, 259
 as ultimate recognition, 55
 Avery's missing prize, 306
 Bell Burnell's missing prize, 11, 12, 49
 Brodie's missing prize, 258
 cermeonies, 4, 147, 179
 Djerassi's missing prize
 Fisher's and Kadanoff's missing prize, 120
 Gajdusek's prize-winning research, 267
 Gamow's missing prize, 92
 Hewish's prize, 49
 Hornykiewicz's missing prize, 275
 Israeli awardees, 67
 Italian awardees, 297
 Klug's missing recognition in the prize for CAT, 284
 Lewis's (G. N.) missing prize, 186
 medal, 85, 86
 Moncada's missing prize, 314
 Ne'eman's missing prize, 65
 physicists of Ioffe's school, 5
 politics and the prize, 211, 212
 rewriting science history, 314
 Russian winners, 9
 Veltman's prize-winning research, 107
 Weissmann's missing prize, 351, 352
nomenclature, 60, 139
non-conservation of parity. *See* parity non-conservation
nonrecurring events, 267
Norrish, R. G. W., 205
Northwestern University, 202, 203
Notgemeinschaft der Deutschen Wissenschaft, 58, 130, 131
nuclear magnetic resonance. *See* NMR
nuclear research, 27, 28, 41, 43, 63, 71, 73, 88, 89, 92, 99, 112
nuclear tests, 31, 32, 99, 201, 308
nuclear weapons, 4, 5, 31–33, 73, 99, 201, 308

nucleic acids, 154, 215, 216, 218, 229, 247, 254, 294
   ribonucleic acid (RNA), 139, 207, 215, 216, 229, 254, 271, 284, 286, 287, 351
   deoxyribonucleic acid. *See* DNA
   sequencing, 213, 215, 216, 270
   protein complexes, 283
Nurse,* Paul, 319–321
Nüsslein-Volhard,* Christiane, 299, 322–324

O'Keefe, Georgia, 127
Oak Ridge National Laboratory, 88, 161
Ochanomizu University, 176
Ochoa, Severo, 286
Ohio State University, 119, 120
Ohio Wesleyan University, 210
Olah,* George A., 193–195
oligonucleotide synthesis, 337
Oliphant,* Mark, 69–71
Open University, 49
Oppenheimer, J. Robert, 99
origin of life, 135, 218, 252, 335, 336
origin of the elements, 70, 162, 219
origin of the universe, 46, 49, 51, 79, 91, 104, 105
Origin of Species, The (Darwin), 83, 188
Osawa, Eiji, 197
Osheroff, Douglas D., 87
Oxford University, 56, 75, 135, 250, 253, 276
ozone, 141, 142, 211

Panofsky,* Wolfgang K. H., 72–74
paradigms, 93, 121, 249, 259
parity and its nonconservation, 28, 88, 89, 97
Parkinson's disease, 257–259, 275–278
Pasteur, Louis, 276
patents, 284, 339, 353, 355
patterns, 40, 51, 91, 130, 177, 254, 267
Pauli, Wolfgang, 97
Pauling,* Linus, 93–95, 186, 196–198, 266, 307, 308, 326, 327, 347
Pavlov, Ivan P., 223
Pedersen, Charles J., 138, 181
penicillin, 136, 151
Penrose,* Roger, 75–77

Penzias,* Arno A., 78–80, 92
Periodic Table, 51, 217
perseverance, 304
personal elements in science, 36, 79, 111, 114, 154, 320, 341, 346
Perutz,* Max F., 325–327
phage. *See* bacteria
pharmaceutical/pharmacological research, 149–151, 243, 269, 256–259, 313, 314
Phillips (company), 303
Phillips, William D., 86
philosophy, 76, 98, 127
photosynthesis, 143–145
Physik Kolloquia. *See* Berlin colloquia
Planck, Max, 42, 57
Platonism, 17
plutonium, 73, 114, 162, 169
plutonium bomb, 32, 114
pneumococcus, 293, 294
poetry, 166, 297, 344, 345
pogroms, 189
Polanyi, Michael, 57, 199
Polanyi,* John C., 178, 199–201
polio research and vaccine, 151, 328–330
Politzer, H. David, 45, 116
Polkinghorne,* John C., 81–83
Pollister, A. W., 293
polyhedra, 159, 160
polysaccharides, 131, 294
Pople,* John A., 202–204
popular science, 111, 205, 206
Porter,* George, 205–206
Post-Capitalist Society (Drucker), 122
Prague Institute of Technology, 207
prayers, 36, 104, 261, 348
prebiotic chemistry, 252, 337
Prelog,* Vladimir, 153, 154, 207–209
Priestley, Joseph, 188
Primakoff, Henry, 133
Princeton University
   Anderson, 6
   Bell Burnell, 9
   Conway, 16
   Gross, 45, 117
   Penzias, 78, 79
   Weinberg, 109
   Wheeler, 112, 113

Wilczek, 116, 117
Wilson (K.), 117
Rowland, 210
Principia (Newton), 121
Principia Mathematica (Russell and Whitehead), 284
prions, 351, 352
Pritchard,* David E., 53–55, 84–86
prognosticating
  Anderson, 8
  Dyson, 24
  't Hooft, 51
  Crutzen, 142
  Fukui, 157
  Lehn, 182
  Olah, 194
  Seaborg, 217, 219, 220
  Semenov, 221
  Levi-Montalcini, 297
  Moncada, 314
  Stent, 335, 336
  Varmus, 342
  Varshavsky, 344
  Watson, 347, 348
Prokhorov, Aleksandr M., 102
proteins
  chemistry, 24, 213, 214, 218, 229, 260
  crystallography, 227, 228
  degradation, 213, 214, 260, 273
  functions, 293, 305, 317, 342, 351, 352
  membranes, 144
  sequencing, 145, 213, 215, 216
  structures, 24, 144, 283, 284, 325, 326
  synthesis, 190, 218, 228, 229, 264
Protestant Churches, 188
Prusiner, Stanley B., 352
psychiatrists, 259, 276, 298, 318
psychology, 127, 156, 218, 291, 318
publishing, 5, 44, 156, 163, 268, 290, 317, 339, 341, 345
Pugwash (movement), 200
pulsars, 9–12, 48, 49
Purcell, Edward M., 88, 89
Purdue University, 125, 245

quantum theory
  chemistry, 142, 158, 159, 197, 198, 203
  mechanics, 7, 8, 27, 46, 76, 98, 105, 121
  physics, 17, 50, 76, 102, 106, 113
quarks, 26, 51, 82
quasars, 9
quasicrystals, 93–95, 197

Rabi, Isidor I., 87, 113, 134
radiation implosion, 31, 32
radioactivity, 223, 271
radioimmunoassay. *See* RIA
Ramakrishnan, Venkatraman, 227
Ramsey,* Norman F., 55, 87–89
Rayleigh, Lord, Strutt, John W., 14
Relativity (Einstein), 26
relativity theory, 105, 344
religion and science
  Anderson, 7
  Arber, 240
  Ciechanover, 261, 262
  Crick, 265
  Mason, 188
  Moncada, 315
  Ne'eman, 67
  Oliphant, 69
  Penrose, 76
  Polkinghorne, 82, 83
  Townes, 103, 104
  Varshavsky, 344
  Weinberg, 111
renormalization, 66, 117
reputation, 118, 130, 200, 290
reserpine, 257, 258, 276
responsibility of scientists, 73, 82, 200, 220, 317
retroviruses, 340, 351
reverse genetics, 351
RIA (radioimmunoassay), 354, 355
ribosome, 227–229, 254
Rich, Alexander, 207
Rickman Godlee Lecture (by Crick), 264
Right to Life movement, 348
RNA. *See* nucleic acids
RNA tie club, 92
Robbins,* Frederick C., 328–330
Roberts, Louis, 88
Robinson, Robert, 135, 208
Rockefeller Foundation, 315

Rockefeller Institute. *See* Rockefeller University
Rockefeller Mauzé, Abby, 21
Rockefeller University, 190, 191, 242, 292, 294, 305, 306, 319
Rollins College, 138
Rose, Irwin A., 260, 273
Rose, Leonard, 287
Rowland,* F. Sherwood, 141, 210–212
Royal Institution (London), 70, 206
Rubens, Peter Paul, 209
Rubin,* Vera C., 90–92
Rudolf II, 188
Russell, Bertrand, 284
Rutherford, Ernest, 43, 69, 70, 71, 137
Ruzicka, Leopold, 153, 208, 209
Ryan, Francis, 294
Ryle, Martin, 48, 49

Sabin, Albert B., 328, 329
Saga of the Nerve Growth Factor, The (Levi-Montalcini), 297
Salam, Abdus, 109
Salk Institute, 253, 264, 337
Salk, Jonas E., 328, 329
Sanger Institute, 337
Sanger,* Frederick, 213–216, 270, 326
Schmidt, Helmut, 67
Schmidt-Ott, Friedrich, 131
Schrödinger equation, 110
Schrödinger, Erwin, 42, 57, 121, 254, 255
Schützenburger, Marcel-Paul, 60
science and art, 320
science and authority, 187–189
science education, 170
Science for Agriculture: A Long-Term Perspective (Huffman and Evenson), 122
scientific cooperation, 4, 5, 156, 263
scrapie, 267, 351, 352
Scripps Florida, 351
SDI. *See* Strategic Defense Initiative
Seaborg,* Glenn T., 161–163, 217–220
Seaborgium, 161, 217
Second Coming, The (Yeats), 297
Segré, Emilio G., 28
self-reproduction, 254

Semenov,* Nikolai N., 4, 221–223
semiconductors, 3, 5
serotonin, 257, 258
SETI, 10
Shalnikov, Aleksandr I., 221
Shechtman,* Dan, 93–95, 197
Shoenberg, David, 43
Siemens (company), 303
Sigler, Paul B., 228
signal transduction, 256, 275
Simon, Franz (Francis), 57, 58
Skou,* Jens Chr., 331–333
Smalley, Richard E., 197
Smith, Hamilton O., 239
Smithies, Oliver, 351
smoking, 145, 251
social issues, 21, 22, 73, 91, 111, 122, 148, 180, 181, 187–189, 262, 264
social sciences, 121, 122, 155, 156, 219, 220
solar energy, 219
solar system, 52, 111, 163
Soviet science, 4, 5, 35, 97, 101, 159, 221–223, 300
space research, 80, 231, 232, 251, 252, 232, 233
Spanish Inquisition, 189
spectroscopies, 62, 142, 144, 194, 234, 290
Spinoza Institute, 50
Sputnik, 4, 73, 97, 223
Stahl, Franklin, 307
Stalin, Iosif V., and his regime, 34, 35, 100, 101, 300, 347
Stanford Linear Accelerator Center (SLAC), 72, 107
Stanford University
 Beadle, 294
 Berg, 247
 Djerassi, 146
 Friedman, 26, 29
 Hofstadter, 29
 Kornberg (A.), 286
 Kornberg (K.), 288
 Lederberg, 292
 McCarty, 305
 Panofsky, 72
 Pauling, 94

Ryan, 294
Shechtman, 94
Tatum, 294
Taylor, 26
Teller, 99, 100
Zare, 230
staphylococcus, 136, 338
State University of New York, 290
Steitz, Thomas A., 227
Stent,* Gunther S., 246, 293, 334–336
stereochemistry, 194, 208, 327
Stieglitz, Alfred, 127
Stieglitz, Julius, 127
Strasshof, Austria, 272–274
Strassmann, Fritz, 162
Strategic Defense Initiative (SDI), 7, 99
Straus, Eugene, 355
strong interactions, 46, 116, 117
structure of DNA. See double helix
stubbornness, 304
Sturkey, Lorenzo, 283
Stuyvesant High School, 166
Sulston,* John E., 253, 337–339
superacids, 194
supercollider, 110
superconductivity, 35, 37, 76, 77
support of science, 97, 218, 219, 332, 341
supramolecular chemistry, 182
Swiss Federal Institute of Technology, Zurich. See ETH
Sydney University, 135
symmetry
  approximate, 145
  asymmetry, 28, 145
  broken, 7, 8, 110
  concept, 16, 88, 110, 117, 145, 197
  fivefold, 197
  icosahedral, 197
  in quantum mechanics, 7
  molecular orbital, 166
  of elementary particles, 28
  of photosynthetic reaction center, 144, 145
  supersymmetry, 66, 110
  twofold, 144
Synge, R. L. M., 214, 215
Syntex (company), 146

Szilard, Leo, 42–44, 57
Sztehlo, Gabor, 193

taboo topics, 249, 318, 347
Talmud, 67, 261, 262, 344
Tamm, Igor E., 34
Tatum, Edward L., 294
Taylor, Joseph H., 12
Taylor, Richard E., 26
teaching of science, 168, 170
Technical University Munich, 62
Technical University Wroclaw, 57
Technion (Israel Institute of Technology), 66, 67, 93, 260, 272
Telegdi,* Valentine L., 28, 96–98
Teller,* Edward, 7, 31, 32, 99–101, 114
Teller-Ulam model, 31, 32
Temin, Howard M., 242
terrorism, 30, 73, 308
tetranucleotide hypothesis, 294
Theory of Everything, 51
thermonuclear explosion, 31
tissue culture, 136, 268, 330
Tisza, Laszlo, 100, 101
tobacco mosaic virus (TMV), 283
Tödliche Wissenschaft (Müller-Hill), 316, 317
Tokyo University, 175–177
Tonegawa, Susumu, 176
totalitarianism, 35
Townes,* Charles H., 87, 102–105
toxicology, 211
transplants, 150, 151
transuranium elements, 162
Trinity College (Cambridge, UK), 81, 202, 204
tritium, 69, 70
Tuppy, Hans, 215
two cultures, 146, 148

ubiquitin, 260, 273, 343
UCLA. See University of California
Udenfriend, Sidney, 258
Ulam, Stanislaw, 31, 32
ultraviolet radiation, 246
uncertainties, 105, 348
Union College, Schenectady, NY, 250
United Nations, 211, 212, 262

universe, 10, 52, 54, 90, 103–105, 162, 241
   origin of, 46, 49, 51, 79, 91, 104, 105
   structure of, 40, 83
   laws of, 46, 83, 87, 103, 113, 254
   history of, 83, 91, 104, 241
Universidad de Buenos Aires, 310
Université Louis Pasteur, 181
University College London
University of Aarhus, 331, 332
University of Alexandria, 234, 235
University of Auckland, 16
University of Basel, 239
University of Berlin, 41, 42
University of Birmingham (UK), 319
University of California,
   Berkeley, 38, 45, 102, 109, 178, 217
   Irvine, 210, 211
   Los Angeles, 138, 190
   San Francisco, 272, 287, 340, 341
   Santa Barbara, 3, 45, 116, 117
University of Cape Town, 282
University of Chicago
   Brown, 125, 127
   Dresselhaus, 19
   Fermi, 27
   Freund, 117
   Friedman, 26, 27
   Garwin, 30, 33
   Hutchins, 127
   Karle (I.), 168, 169
   Karle (J.), 171
   Lee (Y. T.), 178
   Rowland, 210
   Stieglitz, 127
   Telegdi, 96
   Varshavsky, 346
   Westheimer, 224
   Wilczek, 116, 117
University of Colorado, 230
University of Copenhagen, 331, 334
University of East Anglia, 319
University of El Salvador, 313
University of Geneva, 239
University of Gothenburg, 256, 257
University of Illinois, Urbana, 41, 44, 289, 290, 334, 353, 354
University of Kansas, 210
University of Kentucky, 184, 185

University of London, 66, 75, 187, 263, 282, 303, 313, 314
University of Lund, 256
University of Michigan, 38, 39, 106, 168, 169, 171, 173, 174
University of Milton Keynes, 49
University of Minnesota, 299
University of Missouri, 329
University of Nebraska, Lincoln, 138
University of Nottingham, 289, 302, 303
University of Paris, 13
University of Pennsylvania, 132, 234–236
University of Pittsburgh, 290
University of Rochester, 266, 286
University of Sheffield, 206
University of St. Petersburg (Petrograd, Leningrad), 222
University of Southern California, 193
University of Texas, Austin, 109, 112
University of Texas, Dallas, 144
University of Toronto, 199, 200
University of Tübingen, 322
University of Turin, 296
University of Utrecht, 50, 106, 208
University of Vienna, 275
University of Wisconsin, 146, 292
University of Witwatersrand, 253, 282
University of Zagreb, 207
University of Zurich, 350
uranium, 73, 114, 162
Urey, Harold C., 132–134

vaccines, 151, 250, 251, 328
Van Vleck, John H., 6
Vanderbilt University, 295
Varmus,* Harold E., 340–342
Varshavsky,* Alex, 343–345
Vassar College, 90
Veltman,* Martinus, J. G., 50, 106–108
Venter, J. Craig, 338
Vernon-Jones, V. S., 43
Vienna University, 129
Vietnam War, 308, 309
virology, 268, 329
Virus and Rickettsia Laboratory (US Army), 329
viruses, 218, 268, 329, 351
   arboviruses, 268

cancer, 242, 248, 341, 342
causing hemorrhagic fever, 268
chickenpox, 330
hantavirus, 268
hepatitis, 250, 268, 329
herpes simplex, 268
influenza, 268
polio, 328, 330
retroviruses, 351
slow, 267
structure, 282–284
synthesis, 279
tobacco mosaic, 283
unconventional, 267, 351, 352
vaccines, 250
Venezuelan equine encephalitis, 268
vitamins, 94, 152, 153
von Neumann, John, 32, 254
von Weizsäcker, Carl Friedrich, 100

Walter and Eliza Hall Institute, 268
war crimes, 180
Warburg, Otto, 189
Washington University, 132, 247, 286
Watson,* James D., 346–349
  Benzer, 246
  Chargaff, 130
  Crick, 263, 265, 346
  Franklin, 347
  Hopkins, 348
  Kornberg, 287
  Kuroda, 176
  Lederberg, 293
  Luria, 346
  McCarty, 305
  Müller-Hill, 317, 318
  Pauling, 347
  Perutz, 326
  Seaborg, 218
  Stent, 335
  Wilkins, 346
  Wilson, 121, 122
  Yonath, 229
Wayne State University, 125
weak interactions, 28, 50, 63, 64, 97, 106, 107, 117, 182
weapons, 4, 5, 73, 99, 201, 307, 308, 332
Weber, Klaus, 317

Weigle, Jean-Jacques, 246
Weil-Malherbe, H., 276
Weinberg,* Steven, 6–8, 109–111
Weinrich, Marcel, 28
Weissbach, Herbert, 258
Weissmann,* Charles, 350–352
Weizmann Institute, 67, 227, 228
Wellcome Research Laboratories, 313
Wellcome, Henry, 150
Weller, Thomas H., 328–330
Wesleyan University, 86
Western Reserve University, 247
Westheimer Committee, 225
Westheimer,* Frank H., 224–226
What Is Life? (Schrödinger), 254, 255
What Mad Pursuit (Crick), 344
What Mad Pursuit (Keats), 344
Wheeler,* John A., 112–115
Whitehead, A. N., 284
Wieman, Carl E., 53
Wieschaus, Eric F., 300, 322
Wiesner, Jerome, 22
Wigner, Eugene P., 57, 83
Wilczek,* Frank, 46, 116–118
Wilde, Oscar, 25
Wilhelm IV of Hesse, 188
Wilkins, Maurice H. F., 176, 263, 346
Wilson, E. Bright, 122
Wilson, Edward O., 8
Wilson,* Kenneth G., 117, 119–122
Wilson, Robert W., 78, 79, 92
Wilson, Vivian, 69
Wittig, Georg, 125
Wittmann, H. G., 228
Wolfrum, Jürgen, 85
Wollman, Élie, 246, 280, 335
women in science issues
  Brechignac, 13–15
  Dresselhaus, 19–22
  Bell Burnell, 49
  Rubin, 90
  Cohn, 132
  Elion, 151
  Hoffman, 161
  Karle (I.), 170
  Kuroda, 177
  Nüsslein-Volhard, 323, 324
  Hopkins, 348

women in science issues (*Cont.*)
   Watson, 348, 349
   Yalow, 354, 355
Women, Nazis, and Molecular Biology (Stent), 334
Woods Hole, MA, 272, 273, 307
Woodward, Robert B., 165
Woodward-Hoffmann rules, 165
Wooley, Wayne, 190, 191
World Meteorological Organization (WMO), 210, 212
World War I (First World War), 42, 58, 315
World War II (Second World War), 7, 59, 69, 87, 96, 170, 176, 179, 180, 224, 263, 279
Wright Junior College, 127
Wu, Chien-Shiung, 28, 88

X-ray crystallography, 144, 172, 176, 283, 293, 326, 327
X-ray computer assisted tomography (CAT), 284
X-ray treatement, 190, 192
X-rays, 32, 49

Yale University, 130, 292
Yalow, Aaron, 354, 355
Yalow,* Rosalyn, 21, 353–355
Yang, Chen Ning, 28, 88, 107
Yeats, W. B., 295, 297
Yom Kippur, 261
Yonath,* Ada, 227–229

Zare,* Richard, 230–233
Zeldovich, Yakov B., 4
Zewail,* Ahmed, 234–236